In Celebration of

K C Hines

In Celebration of
KC Hines

editors

Bruce H J McKellar · Ken Amos

University of Melbourne, Australia

World Scientific

NEW JERSEY · LONDON · SINGAPORE · BEIJING · SHANGHAI · HONG KONG · TAIPEI · CHENNAI

Published by

World Scientific Publishing Co. Pte. Ltd.

5 Toh Tuck Link, Singapore 596224

USA office: 27 Warren Street, Suite 401-402, Hackensack, NJ 07601

UK office: 57 Shelton Street, Covent Garden, London WC2H 9HE

British Library Cataloguing-in-Publication Data
A catalogue record for this book is available from the British Library.

IN CELEBRATION OF K C HINES

ISBN-13 978-981-4293-65-5
ISBN-10 981-4293-65-2

Kenneth Charles Hines (1926–2005)

Preface

Ken Hines came to Melbourne University as an undergraduate in 1944, was appointed a senior lecturer at the University in 1960, and promoted to reader in 1966. He retired in 1991 but continued to do research in physics until his death on 23rd February 2005. On 18th March 2005 several of Ken's colleagues participated in a memorial conference in his honour. The papers delivered at this conference form the core of this book, but there are other chapters written by some who could not be present, but wished to contribute to this celebration of Ken's career.

Ken was a talented theoretical physicist, with a great gift for teaching and for supervising, guiding and mentoring students. Notable in his CV is the list of students who did a MSc with Ken, and then, with Ken's encouragement, went on to do their PhD elsewhere. These students, and his own PhD students went on to establish noteworthy careers in Australia and overseas. It is a great pleasure that so many of them have been able to contribute to this volume, dedicated to the celebration of Ken's life and his physics.

After his honours degree in cosmic ray physics, and a time doing cosmic ray research on Macquarie Island, he completed his PhD in theoretical physics and then worked at Harwell on fast breeder reactors. This research was a part of the U.K. development program on such reactors, leading to their construction in the late 1950s and experimental and commercial use subsequently. He then returned to the Australian Atomic Energy Commission's research establishment at Lucas Heights working on reactor physics. He continued these studies on reactor physics for a brief period after coming to Melbourne, and then moved into plasma physics, an area of interest he maintained in many ways throughout his career. His interest moved from laboratory plasmas to astrophysical plasmas, including some unusual ones, such as particle-antiparticle plasmas. This was a subject which was of great astrophysical interest for a time, when it was realised by Alfvén that such plasmas would form in the boundary region between matter and anti-matter galaxies in the universe, and the search for the annihilation radiation from them was actively pursed. At about the time he was working on this problem, Ken arranged a visit by Alfvén to Melbourne. In his later work, Ken turned to a detailed study of the properties of tachyons, and to the study of the plasma physics involved in accretion on black holes.

Hannes Alfvén was one of Ken's many international contacts who were able to visit Australia. Another was Dirk ter Haar of Oxford, who came to be a friend of many of us in Melbourne. Ken's last international collaboration was with a former Melbourne student, Fulvio Melia of the University of Arizona. Fulvio has contributed a chapter to this book which provides an account of the results of that collaboration. All of his international friends enriched our lives in the School of Physics at Melbourne University.

For many years Ken played a significant role in two of the ongoing conference series in Australian physics, the plasma physics conferences sponsored by the Australian Institute of Nuclear Science and Engineering and held at Lucas Heights every second year, and the Australian Mathematical Society's Applied Mathematics Conferences. Ken's contribution to these conferences do not appear in his publications list, as proceedings were not published, but his formal contributions were often overshadowed by his contribution to the discussion, during conference sessions, at the dinner table and late into the evening.

As you will gather from the obituary which appeared in the Melbourne Age which is part of Chapter 1, Ken's skills were not confined to physics, but extended to literature, the languages and culture of Germany and Italy, and to music. He was a well-known face of the School of Physics on the campus, and spent many years on the executive of the Staff Association. He also spent 6 years as the Deputy Head of the School of Physics.

By his teaching, research and fellowship he enriched the life of the Melbourne University School of Physics, and the lives of those who worked in it and passed through it. Ken's contributions should be celebrated, and we are pleased to be able to do so with this volume of contributions by his friends.

Ken Amos
Bruce McKellar

Contents

Chapter 1

Concerning Ken Hines ...

1.1. Obituary published in the 'Age' newspaper

KENNETH CHARLES HINES:
Physicist, Antarctic expeditioner
22-9-1926–23-2-2005

From a hole in the ice to black holes in the sky
Philip Jones

Ken 'Tonky' Hines, an intrepid man of science and action who nearly lost his life while skiing in the antarctic with a fellow expeditioner who ended up drowning, has died in Melbourne. Hines was 78.

All his life, Hines concerned himself with the lives of others, and he was intellectually cultivated — his interests ranged from literature to music. He was a linguist, versed in French and Italian and their literature, Latin and German, with a special interest in the poetry and drama of Johann Schiller. His teacher was the inspiring Anita Rogers.

Hines ancestors migrated from various parts of the British isles in the late 19th century. His great grandfather, Samuel Hartley Roberts, a decorator of public buildings and one-time mayor of Richmond, (in Melbourne, Victoria) died after a fall from the great dome of the Exhibition building.

Hines parents were married in St. Paul's cathedral — his father was a grain merchant — and the family settled in Kew. He studied as Trinity Grammar. Hines was a spectacular ruckman in the school's (Australian rules) football team — the game was a passion for at least four generations of Hineses, who were all Demon (Melbourne football club) supporters. But he was bored by cricket; Hines claimed that whenever he was forced to 'field', he invariably missed the ball because he was reading a book! In 1942 he passed the school leaving examination in eight subjects and was dux of his form. In his matriculation year he obtained honours in German, physics, chemistry, and mathematics — and always paid tribute to his beloved german teacher, Rogers, who first fired his enthusiasm for the language and literature of that country. In later life he served on Trinity Grammar's council.

Hines went on, in 1944, to study science at Melbourne University, majoring in physics. He had just begun an MSc in cosmic ray physics in 1947 when he and his friend, Leigh Speedy, were recruited by Dr. Philip Law, the original leader of Australia's Antarctic expeditions, to join what became known as "the cosmic ray group" on Macquarie island. The aim was to compare the effects of southern latitude rays with similar research on Heard Island.

Geiger counters to measure cosmic rays were primitive compared with today's and were tested on the snow fields of Mount Hotham. Hines recalled that down on Macquarie, diesel power generators were erratic in their supply of electricity and had to be "greased" with toothpaste. Hines and his colleagues were confined to their hut for days at a time by snow. "It rained every day of the year that we were there", he later recalled,

On this first expedition little was known about the terrain, and which activities were and were not safe. On July 4th, 1948, Hines and his friend, a diesel engineer named Charlie Scroble, decided to go skiing. They took their skis and climbed up to a plateau. Here they skied over a frozen lake. Their exhilaration came to a sudden end when, in the middle of the lake, the ice broke and they plunged into freezing water. Hines managed to remove his skies and used one as a lever to gradually edge himself to the shore. He wrote later: "We knew the nearest human beings were several miles away and separated from us by a thousand-foot climb and it was no use shouting for help". Despite Hines' constant urging, Scroble, a big, heavy man, could not make the distance. Finally he sank and drowned. Hines, who was exhausted by his own struggle to get out, had to rest for several hours before he managed to make it back to base. Scroble's body was recovered in the spring and he was buried beside the lake.

While waiting to start his studies for a masters degree in the academic year of 1950, Hines attended German classes at Melbourne University, where he was delighted to find his old tutor, Rogers. In this class he met Suzanne Baer, with whom he fell instantly in love. "I knew immediately this was the one woman I had to marry." It was a marriage, as they say, made in heaven.

Hines changed his study from experimental to theoretical physics and gained an MSc and then a doctorate. He made significant contributions over

Fig. 1.1. The team on Macquarie Island, 1948.

his academic years at his alma mater. His first professional appointment was in England at the Atomic Energy Research Establishment. Later, he worked at the Australian Atomic Energy Commission and at the Oak Ridge National Laboratory in the United States.

At his funeral, no less than six former colleagues paid tribute to his work and to the generosity of his spirit.

After his retirement in 1991, he spent the rest of the decade as a senior research associate working on the astrophysics of black holes.

Hines was a bon vivant and belonged to a luncheon club that included friends from the university and the Antarctic expeditions.

His family will always remember his many fine qualities, perhaps none more than his instinctive sympathy and love for children. He also possessed a lifelong affection for elephants.

Hines leaves his wife, Suzanne, and his three sons, David, Toby, and Michael.

1.2. Curriculum vitae: Kenneth Charles Hines

(1) Personal details:
- Date of birth 22nd Sept., 1926
- Place of birth Melbourne, Australia

(2) Academic record:
- Dux of class Trinity Grammar School, 1942
  School leaving certificate with honours in
  German, Physics, Chemistry, and Mathematics
- B. Sc. (hons) University of Melbourne, 1947
- M. Sc. University of Melbourne, 1949
- Ph. D. University of Melbourne, 1954

(3) Employment:
- 1991–2005 (rtd.) Principal fellow, University of Melbourne
- 1966–1991 Reader in physics, University of Melbourne
- 1964–1966 Science correspondent of the 'Age' newspaper
- 1966–1967 Research associate, Thermonuclear division,
  Oak Ridge National Lab., Oak Ridge, Tennessee, U.S.A.
- 1960–1966 Senior lecturer in physics, University of Melbourne
- 1956–1959 Research officer/senior research officer,
  Australian Atomic Energy Commission
- 1954–1955 Senior scientific officer,
  Atomic Energy Research Establishment, Harwell, England
- 1946–1949 Cosmic ray physicist,
  Australian National Antarctic Research Expedition
  on Macquarie Island

(4) Review of research:
- Reactor physics:
  (a) Harwell 1955–1956: Research on the physics of fast breeder reactors. Calculation of the critical enrichment of the Harwell fast reactor, ZEUS.
  (b) Harwell and Lucas Heights (Australia) 1957–1959: Research on the behaviour of thermal reactors, development of a new method for calculating energy distributions in common moderators.
  (c) Lucas Heights and University of Melbourne 1960–1963: Research on the slowing down on neutrons in moderators with

an enhancement mechanism (Be, BeO).

- Plasma physics:
  Energy loss rate of particles in a plasma — relativistic and non-relativistic plasmas, quantum and non-quantum cases. Calculations of the dielectric functions for quantum relativistic plasmas, 1964–1979 University of Melbourne.
- Astrophysics:
  research on plasma astrophysics, especially on the properties of compact X-ray sources, 1970–2004 University of Melbourne.

(5) Supervision of higher degree students:

- Ph. D. students

  (a) V. Buzzi: Relativistic two fluid plasmas in the vicinity of a Schwarzchild black hole (1994)

  (b) B. K. Smith: Systematics of the electroweak plasma at finite temperature (1994)

  (c) R. L. Dawe: The physics of faster than light objects (1990)

  (d) N. S. Witte: The response theory of the magnetised and unmagnetised, finite temperature, spin-0 pair plasma and vacuum (1989)

  (e) V. Kowalenko: (partial supervision) Relativistic quantum plasmas (1981)

  (f) A. R. Carr: Lagrangian analysis of non-linear wave-wave interactions in bounded plasmas (1979)

  (g) P. Cadusch: A Lagrangian including particle correlations for warm plasmas (1977)

  (h) R. A. O'Sullivan: Relativistic theory of el;ectromagnetic susceptibility and its applications to plasmas (1974)

  (i) R. A. Lee: Self energy functions and the nuclear optical model (1974)

  (j) B. C. H. Wendlandt: Studies in plasma physics (1970)

  (k) N. E. Frankel: Studies in statistical physics (1967)

  (l) J. C. Herzel: Thermodynamic Green functions (1966)

- M. Sc. students
  1963 – T. H. Axford
  1965 – J. B. Marsh
  1967 – A. J. R. Prentice, R. L. Dewar, G. G. Lister
  1969 – J. F. Dobson
  1971 – G. Briscoe, E. Toime

1973 – L. T. Cohen
1974 – R. D. Burrows
1984 – A. M. Dimits
1985 – M. Baring

- B. Sc. Hons. students
1987 – C. Lidman
1988 – G. Guest, H. Lee
1989 – D. Henshaw
1990 – J. Gunning, W. Rowlands

(6) <u>Publications:</u>

- As author/coauthor:

(a) K. C. Hines,
The straggling of electrons below the critical energy,
Aust. J. Sci. Res. **4**, 450 (1951).

(b) J. R. Bird and K. C. Hines,
The multiple scattering of protons in nuclear emulsions,
Aust. J. Sci. Res. **7**, 586 (1954).

(c) K. C. Hines and J. P. Pollard,
Slowing down of neutrons in Be and BeO,
J. Nucl. Eng. **16**, 71 (1962).

(d) T. H. Axford, K. C. Hines, and J. P. Pollard,
Neutron slowing down spectra ion Be and BeO,
J. Nucl. Eng. **18**, 131 (1964).

(e) N. E. Frankel, K. C. Hines, and R. L. Dewar,
Energy loss due to binary collisions in a relativistic plasma,
Phys. Rev. A **20**, 2120 (1979).

(f) V. Kowalenko, N. E. Frankel, and K. C. Hines,
Response theory of particle-antiparticle plasmas,
Phys. Rep. C **126**, 109 (1985).

(g) N. E. Frankel, K. C. Hines, and V. Kowalenko,
Relativistic boson-antiboson plasma,
Lasers and Electronic Beams, **3**, 251 (1985).

(h) N. S. Witte, R. L. Dawe, and K. C. Hines,
Relativistic charged bosons in a magnetic field 1: Wave functions and matrix elements,
J. Math. Phys. **28**, 1864 (1987).

(i) V. Buzzi, K. C. Hines, R. A. Treumann,

Relativistic two plasmas in the vicinity of a Schwarzschild black hole I,
Phys. Rev. D **51**, 6663 (1995).

(j) V. Buzzi, K. C. Hines, R. A. Treumann,
Relativistic two plasmas in the vicinity of a Schwarzschild black hole II,
Phys. Rev. D **51**, 6677 (1995).

(k) V. Buzzi and K. C. Hines,
Relativistic plasmas near a Schwarzschild black hole,
Phys. Rev. D **51**, 6692 (1995).

(l) B. J. K. Smith, N. S. Witte, and K. C. Hines,
Systematics of the electroweak plasma at finite temperature I,
Aust. J. Phys. **48**, 739 (1995).

(m) B. J. K. Smith, N. S. Witte, and K. C. Hines,
Systematics of the electroweak plasma at finite temperature II,
Aust. J. Phys. **48**, 775 (1995).

(n) R. L. Dawe and K. C. Hines,
The physics of tachyons IV,
Aust. J. Phys. **51**, 477 (1998).

- Under supervision:

(a) N. E. Frankel,
On the Fokker-Planck coefficients for an inverse square law force,
Phys. Lett. **1**, 315 (1965).

(b) N. E. Frankel,
Dominant and non-dominant terms in the energy transfer equations for a fully ionised plasma,
Plasma Phys. **C7**, 225 (1965).

(c) A. J. R. Prentice,
Collective energy loss in relativistic plasmas,
Plasma Phys. **9**, 433 (1967).

(d) J. C. Herzel,
Green's functions and double-time distribution functions in classical statistical mechanics,
J. Math. Phys. **8**, 1650 (1967).

(e) J. C. Herzel,
Many-time Green functions and the resolvent operator,
Phys. Letts. **A27**, 654 (1968).

(f) J. C. Herzel,
Classical Green functions in response theory,
J. Math. Phys. **11**, 741 (1970).

(g) R. A. O'Sullivan and H. Derfler,
Relativistic theory of electromagnetic susceptibility and its applications to plasmas,
Phys. Rev. A**8**, 2645 (1973).

• Conference papers:

(a) K. C. Hines,
Single particle and collective effects for the test particle problem in a plasma,
Bull. Am. Phys. Soc. **12**, 26 (1967).

(b) K. C. Hines and D. J. Sigmar,
Energy loss of charged particles in a plasma,
Conf. on 'Phenomena in Ionised Gases', Oxford, 295 (1967).

(c) K. C. Hines,
Self-consistent electric field for a Vlasov plasma,
Risö report **250**, Danish A. E. C. in Proc. 3rd Int. Conf. on Quiescent Plasmas', 365 (1971).

(d) K. C. Hines and D. F. Hines,
A unified relativistic theory of test particle energy loss in a plasma, Conf. on Phenomena in Ionized Gases, Berlin (D.D.R), 734 (1977).

(e) N. E. Frankel, K. C. Hines, and R. D. B. Speirs,
Intermediate quantum plasma in the laser driven fusion regime,
Conf. on Phenomena in Ionized Gases, Grenoble, 513 (1979).

(f) N. E. Frankel, K. C. Hines, and R. D. B. Speirs,
Dielectric response and energy loss for an intermediate quantum plasma,
J. de Physique **C7**, 513 (1979).

(g) N. E. Frankel, K. C. Hines, and V. Kowalenko,
Dielectric response of particle-antiparticle plasmas in a magnetic field,
Proc. Spring College of Fusion Physics (Ed. B. McNamara), Trieste AES-SMR **82**, 353 (1981).

(h) A. M. Dimits and K. C. Hines,
Relaxation of a fast ion in a plasma; a relativistic treatment,
Conf. on Phenomena in Ionized Gases, Dusseldorf, 38 (1983).

(i) V. Buzzi, N. E. Frankel, K. C. Hines, H. H. Lee, and N. S. Witte,
Charged boson and fermion pair plasmas at finite temperature,
Conf. on Phenomena in Ionized Gases, Belgrade, 270 (1989).

- Unpublished reports:

(a) K. C. Hines, *Secret report on fast reactor theory,*
A. E. R. E classified report (1955).

(b) K. C. Hines and J. Codd, *Secret report on fast reactor theory,*
A. E. R. E classified report (1956).

(c) K. C. Hines and J. Egerton, *The approach to criticality of ZEUS,*
A. E. R. E RP/M 81, 1 (1956).

(d) K. C. Hines and J. Egerton, *Solution of the pile equations of a bare cylindrical system of hexagonal cross section,*
A. E. R. E RP 2248, 1 (1957).

(e) K. C. Hines, *Energy and lethargy distribution of neutrons slowing down in graphite,*
A. E. E. C. E/36, 1 (1959).

(f) E. Duncan, K. C. Hines, and J. P. Pollard, *Slowing down spectra of neutrons in heavy water and light water mixtures,*
A. E. E. C. E/78, 1 (1961).

(g) K. C. Hines, G. Joyce, and D. J. Sigmar, *Exchange energy of a test particle with a plasma,*
ORNL: Thermonuclear division semi-annual report, Oct. (1967).

1.3. Some short stories about Ken

There are many stories about Ken Hines that his friends recall and cherish. Here are a few of their favourites.

1.3.1. *Roger Hosking reminisces*

A year or so after I took up my first academic job at Flinders University in early 1966, following my doctorate in Canada and a post-doctoral at the Max Planck Institute in Munich, Graeme Lister became my first PhD student on the recommendation of Ken Hines. My first wife and I shared a table at the reception when Graeme was married shortly afterward, and Ken and 'Suza' Hines instantly became two of our closest personal friends.

Ken's charm and 'mana' were more than enough to persuade the waitress at the reception that we deserved extra helpings and rather more wine at our table. Forever after, I would take every opportunity to join him at the (in)famous Thursday lunch club at a restaurant in Lygon Street or nearby! Our families also met on many occasions over the years, either at the Hines' North Balwyn family home or at the first or second of their beach houses at Moggs Creek near Lorne. The first beach house unfortunately was destroyed by a monstrous bushfire, leaving only melted glass from a few essential provisions Ken kept there!

There were also the annual AINSE (Australian Institute of Nuclear Science and Engineering) plasma physics meetings, which I first attended in early 1967. On one occasion, Ken gave a memorable after dinner speech in which he referred to the Father (Charlie Watson-Munro at Sydney), the Son (Max Brennan, then at Flinders) and the Holy Ghost of plasma physics (he, at Melbourne).

1.3.2. *Ken Amos reminisces*

Julie and I with three young boys were staying for a day or so at the Hines holiday house at Mogg's Creek. Brian, our youngest, was just learning to play the basic recorder, and Ken asked him if he would like to play something with him (Ken on the basso-profundus – an important instrument that one of Ken's sons made a point of saving from the devastation of Ash Wednesday). All the rest of us took to the beach and its delights while Ken and Brian took to a lengthy study of a two recorder piece. In fact that collaboration and education of a nine year old lasted most of the day. On our return when it grew dark, it was announced that after dinner there would be a recital. They simply amazed us and it revealed what patience, teaching ability, and general love of children Ken had.

Later, in Victor Harbour in South Australia, many years ago, Ken, I, and a number of colleagues in theory group attended an Applied Mathematics conference. There, I was faced with Ken's unshakeable belief that, as I was born and raised in the state (albeit that I had been elsewhere for nearly 20 years), I knew every beach worth knowing. Despite my protestations that such was not true, Ken was very persistent. So, on the third morning as I recall, I agreed to set him upon a beach. In swimsuit with towel, Ken came with us in our car we drove off along the same road we had taken to the conference on both preceding two days. A short way out of the town I stopped the car and pointed to a path and sign which read "To the beach". So off Ken toddled while we continued on the drive. About 30 seconds later, Norm Frankel said "Ken, you are a perverse b——! We have passed this place many times and you didn't tell Ken .... Oh hell, you really don't know this place do you! What if ...?" Well "What if ...?" came all too true. The so-called beach was an unmitigated disaster. As we eventually found out, the beach and water was a mire of seaweed and rocks and the water did not get to swimming depth until so far off-shore that the current would have taken our intrepid swimmer back to Victoria. Well Ken struggled back to our motel and took his "swim" in a hip bath of a swimming pool. Needless to say, our return was greeted rather icily. Indeed, it was one time that I can truly say that Ken was speechless. Well for a while anyway. But as he was such a great and kind man, and by monopolising our bottle of scotch, Ken soon forgave me.

1.3.3. *Vic Kowalenko reminisces*

Shortly after completing my Ph.D. thesis on the response theory of particle/anti-particle plasmas in 1982, I managed to secure a research position at Materials Research Laboratories (MRL), now part of DSTO Melbourne, assisted no doubt as I had the unique honour of having had my entire Ph.D. thesis accepted for publication in the prestigious Physics Reports. Ken was instrumental in preparing all for that publication, and as there was an excellent typist at MRL to produce the manuscript I had get approval for Ken to carry out this work at MRL. It was amusing to see Ken moving around MRL as if he were an "old hand" there. In the end he had made friends with my various colleagues working on the plasma armature rail gun. On the day we mailed the final manuscript to Oxford, we celebrated with a fine meal at a nearby Footscray restaurant, for which he offered to pay. There was no doubt that he was proud of his involvement in the publication of the thesis.

A few years after the publication, Ken informed me that his second son's fiance had been visiting relatives and friends in Israel to announce the impending marriage. At a party in her honour she was introduced to someone, who informed her that he was a professor of physics. She informed him that her future father-in-law was also a physicist. He then asked for the name and she replied immediately. She did not expect a response, but after a brief pause, he muttered the Hines in "Kowalenko, Frankel and Hines", the exact order of the authors in the Physics Reports.

1.3.4. *Zwi Barnea reminisces*

Ken Hines loved larrikins and larrikinism and consequently always delighted in telling us the story of how he met Terry Sabine:

Ken met Terry Sabine for the first time accidentally on the ferry from France to England. It was in the fifties when both were working at Harwell. When they arrived in Dover, there was a problem with Terry's car; some document was missing. Terry asked where he could park his car until he arranged the documentation. The customs officials were not helpful and there were no provisions for parking. Terry opened the door of his car (it was not a particularly new one), calmly released the hand-brake, and gave the car a good push. The car lurched forward, fell into the dirty waters of the port, and sank slowly and majestically, much to the horror of the "Pommy" customs officials.

1.3.5. *"Legend's" Thursday lunch club award number four*

The Thursday lunch club membership unanimously gave the Legend award number four to Ken (means beans) Hines for the following event.

We were at a hotel in Fitzroy for a particular, and one-time visit, lunch when Ken started a splendid eulogy to an old and valued colleague in Physics. When Ken eulogised, he did so in very Italian style, so with all the flair of one of the worlds great orators, Ken managed to knock the elbow of a passing waiter who had just cleared our table. The collision sent a sizeable stack of plates with cutlery crashing to the terrazzo floor. It was only a pity that it was not a truly Greek restaurant for the demolition was

Fig. 1.2. One Thursday about 2 p.m. ...

loud and complete. Not phased, though apologetic, Ken resumed the long eulogy at exactly the point of interruption. For many other reasons, that eulogy was never completed.

Legend award number nine also went to Ken Hines. The gentleman of our troupe who always had a warm greeting and some wry remarks about almost any topic one chose to discuss with him. He has departed us leaving for a quieter repose with Newton, Einstein and just about any of those wizards of natural philosophy past. Ken, you have been an inspiration to all of the lunch crew and we'll miss you. Rest in peace our most cherished friend.

1.3.6. *Graeme Lister reminisces*

My early memories of Ken are irrevocably linked to the making of my first hangover. Not that he was any way responsible, but the event occurred when I was a young graduate student, at my first Melbourne Physics Department dinner. I grew up in a family where alcohol was regarded as the road to ruin, and my mother would warn me on each occasion I attended a party to beware of the demon drink. At that stage in my life, I didn't particularly care for the taste of alcohol, but on the occasion in question, my mother's warning led me to question whether there was more to a glass of wine than meets the palate.

It was a "6 for 7" affair, which turned out to be "6 for 8", and I tried all the drinks offered in succession, my unrefined palate finding cream sherry much to its liking. I found myself standing behind Ken in the queue for the buffet. My eyes opened as I watched him pile his plate high with food, and I matched him, chicken breast for chicken breast, ham slice for ham slice. The meal took some consuming, so I washed it down with white wine. I was then extremely thirsty, and attacked the offered beer with gusto. By this stage, I was certainly feeling the effects of my over consumption, and my last memory of Ken that evening, from the corner of my eye, is him approaching the table for a second, equally large helping. I tried to clear my head by walking from the University to Flinders St. Station, and although I have only vague memories of the journey, I distinctly remember falling down the station steps and being helped to my train. I fell asleep on the home journey, waking up a Ferntree Gully, the end of the line. I took a taxi home to face the music, then woke next day swearing never to do it again ... a forlorn promise. Over the next three decades, I raised many a glass of the best wine with Ken, and he left me with many fond memories.

1.3.7. *Bob Dewar reminisces*

Ken Hines was our lecturer in 3rd Year Electromagnetism, where we also learnt Special Relativity, and Ken's presentations of these subjects attracted four of the theoretically inclined members (including me) of my class to him as supervisor for postgraduate research projects. At the time it was standard for those continuing beyond 3rd Year at Melbourne to do at least an MSc, with little concept of BSc Hons as a degree in its own right. Ken started a reading group in statistical physics where we discussed various seminal papers. He got us interested in such topics as the dressed test particle picture, but it is my recollection that he took a sabbatical year at Oak Ridge fairly early during my Masters course. This did not greatly slow research progress, as, with the large cohort from my year plus some older postgrad students we had sufficient critical mass that we could teach each other, thus learning valuable lessons in both independence and cooperation. By the time he returned I had made sufficient progress on my MSc topics that Ken played the role of avuncular senior mentor rather than prescriptive supervisor. I left Melbourne in 1967 to do my PhD at Princeton, but the plasma physics background I had gained at Melbourne stood me in good stead, as did Ken's teaching of the importance of the other things in life, such as good food, conversation and music. On subsequent (all too infrequent) visits to Melbourne I would always call in on Ken and appreciated his warm hospitality, both in Lygon St lunches and at his home.

1.3.8. *Norm Frankel reminisces*

Ken and I were the very best of mates. In the '70s an '80s, something we enjoyed doing was attending the applied mathematics meetings of the Australian Mathematical Society. These were held in beautiful spots like the Barossa Valley and Victor Harbour in South Australia, Jindabyne in the high country of New South Wales, and Broadbeach on the Gold Coast of Queensland. It was a very sunny and warm time in the early 1970's when we participated in the latter one, held in the Broadbeach Hotel only a stones throw from the warm, beautiful, Pacific ocean. The first morning when we awoke around 6 a.m, the sun was streaming in through the open glass doors that led onto the balcony of our room. Stimulated by the sun's warmth and illumination, Ken ushered us out onto the balcony, with its glorious view. We were in our birthday suits. Inspired, Ken stretching his arms up high and his body to full extension, announced in his unique stentorian voice, "GOOD MORNING QUEENSLAND".

A moment later, a soft dulcet voice responded with "Good morning". At just this moment, unbeknown to us, directly below us a lovely young woman had arrived to waitress. Ken ushered us back into our room straight away with a somewhat bemused look on his face. We said nothing, dressed and went down to the large dining room for breakfast. When we got to the entrance to the dining room, an attractive waitress greeted us with "good morning". Yes it was the same lass. We chose our breakfast from the cornucopia of fruits and assorted delights and ate in silence, looking up at times to each other with the gentlest of Cheshire Cat grins. From that time on, "GOOD MORNING QUEENSLAND" was our anthem.

Chapter 2

Structures of Exotic Nuclei from Nucleon-Nucleus Scattering Data

Ken Amos

School of Physics, University of Melbourne
Victoria 3010, Australia
E-mail: amos@physics.unimelb.edu.au

Two theories of nucleon-nucleus scattering have been used to investigate the spectra of the isotopes of Carbon, ranging across the periodic table from proton to neutron drip lines. The first method considered used a multichannel algebraic scattering (MCAS) theory for neutron - ^{12,14,16,18}C interactions as coupled-channel problems. This approach has been used to identify structures of the compound systems, giving information on both sub-threshold (bound) and resonance states. Low energy scattering cross sections are also obtained thereby. The second method is pertinent to analyse scattering data, elastic and inelastic, with results being a test of shell model structure assumed for the target nuclei, or for radioactive ion beams scattering from Hydrogen targets by using inverse kinematics. The method, termed g-folding, has, as input. an effective two-nucleon interaction, one-body density matrix elements of the target states, and single nucleon bound state wave functions. With the latter two elements defined by large space, shell model studies, and the first set as the 'Melbourne' force, cross-section and spin-observable data are well predicted with little need of effective charges when the structure model used is an appropriate description of the spectrum of the nucleus concerned.

2.1. Introduction

I am delighted to pen this article in memory of my colleague and friend of over 30 years, Ken Hines. Ken studied problems in nuclear physics during his years at Harwell and then at the Lucas Heights facility in New South Wales, Australia. But that was many years ago. Nonetheless one of the problems he considered then was how low energy nucleons interact with nuclei, and a most modern version of that is one of the topics of this paper.

An ultimate aim of scattering theory is to probe the structure of nuclei. Data may be taken with electrons, mesons, nucleons, and nuclei as probes, with form factors, cross sections, and spin observables being data. There are diverse reactions one might consider (elastic, inelastic scattering, and particle exchange reactions are some). But almost all theories rely upon elastic scattering information and require, as input, model specifics of the nuclear structure. However, an important requirement in nuclear theory is the (model) construction of an interaction Hamiltonian for the nucleon-nucleus and/or nucleus-nucleus systems, that is realistic for both bound states and scattering (including resonances). To do so is important for many reasons; theoretical consistency, evaluation of Electromagnetic (EM) transition rates, prediction of radiative capture cross sections etc., as such are required input for cluster-like, three-body calculations of core nucleus-N-N, core-core-N, and core-core-core systems. That is especially so when considering systems that are important for current and future experimental researches using radioactive ion beams (RIBs). Physically justified cross sections are also important in making nuclear-data evaluation that are used in applications such as of radiation safety, nuclear medicine, nuclear astrophysics, and nuclear weapon stewardship. Attaining a theory of that kind remains a 'holy grail', and approximate methods are still the only practical means for studies of nuclear structure and reactions.

The real question then is what structure model to use and what reaction theory is appropriate? Hereafter I consider nucleon-nucleus systems but generalise briefly in a study of two hyperon-nucleus cases. The structure of immediate concern then is that of the nucleus in the nucleon-nucleus system of interest.

Consider first what light mass nuclear spectra are in general. At low excitations, there are discrete states while at excitations above any giant resonances, there is a continuum of states. Those aspects of a target effect the choice of scattering theory to be used. For low incident energies, discrete states of the target should play significant roles. This means one needs to solve a coupled-channels problem. The MCAS approach [1] was designed to do that. For higher incident energies, however, the continuum is the important feature of the target and there are so many states of the target to effect a response, that an average scheme may be relevant. The g-folding optical potential model (for elastic scattering) and the distorted wave approximation (DWA) model for inelastic processes works very well in those cases [2].

With low energy situations, the MCAS method gives spectra of compound nuclei formed by the amalgamation of two separate nuclear clusters. Sub-threshold (bound states) if they exist, as well as resonance states (for energies above the two cluster threshold) can be deduced using it. Those resonances can be narrow or broad. Some of the first type can be identified as bound states embedded in the continuum (BSEC) while broad resonances are characteristic of single particle potential resonances. It has been noted in the literature that

"The BSEC phenomena can lead to a change in the level density close to the continuum threshold, so such structures will be highly important in the astrophysical capture and knockout reactions contributing to nuclear synthesis in neutron rich stellar environments".

For scattering at energies above a few tens of MeV per nucleon, the number of target states to be included in a full MCAS calculation makes evaluation with that scheme no longer feasible. However, when the energies coincide with a region of high density of states in the target, an alternative approach, that has proved very successful, is to use the g-folding optical potentials in an analysis of elastic scattering observables, and to use the relative motion wave functions they produce as the distorted waves in a DWA analysis of inelastic scattering data. The underlying two-nucleon (NN) g-matrices in the generation of the optical potentials are complex, medium- and energy-dependent and are also used as the transition operators effecting the inelastic transitions. It is important to note that the g-folding optical potentials are very non-local and that the actual non-locality must be used in evaluations. It is inadequate to approximate the problem using some local 'equivalent' potential. Often, and to this day, phenomenological, local optical model potentials are used to determine the distorted wave functions of relative motion with the argument that a quality fit to the elastic scattering data justifies use of the associated relative motion wave functions. But such fits only require specification of a suitable set of phase shifts which are determined from the asymptotic properties of the solutions. The credibility of distorted wave functions through the volume of the nucleus cannot be assured thereby. Indeed, it has long been known that those wave functions are too large through the nuclear volume due to inadequate representation of nonlocal effects.

In the next two sections, I present elements of the MCAS and g-folding methods along with some results of applications to illustrate the flexibility and use of them. Then in Section 2.4 I employ both methods in a study of the spectra of the isotopes of carbon ranging between the drip lines. Another very recent application of the MCAS method, namely to define spectra of Λ hypernuclei, is reported in Section 2.5, after which brief concluding remarks are given.

2.2. MCAS and nucleon-nucleus systems at low and negative energies

In brief, the MCAS approach is based upon using sturmian functions as a basis set to expand the chosen interaction potentials. Each interaction matrix then has the form of a sum of separable interactions. The analytic properties of the S-matrix from a separable Schrödinger potential gives the means by which a full algebraic solution of the multichannel scattering problem can be realized. All details of the MCAS theory have been published [1] and so only salient features are repeated herein. Consider a coupled-channel system for each allowed scattering spin-parity J^π. With the MCAS method, one solves the Lippmann-Schwinger (LS) integral equations in momentum space,

$$T_{cc'}(p, q; E) = V_{cc'}(p, q) + \frac{2\mu}{\hbar^2} \left[\sum_{c''=1}^{\text{open}} \int_0^\infty \frac{V_{cc''}(p, x)\ T_{c''c'}(x, q; E)}{k_{c''}^2 - x^2 + i\epsilon} x^2\, dx \right.$$
$$\left. - \sum_{c''=1}^{\text{closed}} \int_0^\infty \frac{V_{cc''}(p, x)\ T_{c''c'}(x, q; E)}{h_{c''}^2 + x^2} x^2\, dx \right],$$

where the index c denotes the quantum numbers that identify each channel uniquely. Such requires specification of potential matrices $V_{cc'}^{(J^\pi)}(p, q)$. The open and closed channels have channel wave numbers k_c and h_c for $E > \epsilon_c$ and $E < \epsilon_c$ respectively. μ is the reduced mass. Solutions of these LS equations are sought using expansions of the potential matrix elements in (finite) sums of energy-independent separable terms,

$$V_{cc'}(p, q) \sim \sum_{n=1}^N \chi_{cn}(p)\, \eta_n^{-1}\, \chi_{c'n}(q).$$

The key to the method is the choice of the expansion form factors $\chi_{cn}(q)$. Optimal ones have been found from the sturmian functions that are determined from the actual (coordinate space) model interaction $V_{cc'}(r)$ initially chosen to describe the coupled-channel problem.

The link between the multichannel T- and the scattering (S) matrices involves a Green's function matrix,

$$(G_0)_{nn'} = \frac{2\mu}{\hbar^2} \int_0^\infty \left[\sum_{c=1}^{\text{open}} \frac{\chi_{cn}(x)\,\chi_{cn'}(x)}{k_c^2 - x^2 + i\epsilon} - \sum_{c=1}^{\text{closed}} \frac{\chi_{cn}(x)\,\chi_{cn'}(x)}{h_c^2 + x^2} \right] x^2\, dx,$$

where $(\eta)_{nn'}$ is a diagonal eigenvalue matrix $(\eta_n\,\delta_{nn'})$.

The bound states of the system are determined from a matrix determinant for energy $E < 0$. They are found from the zeros of $\{|\boldsymbol{\eta} - \mathbf{G}_0|\}$ when all channels in the above equation for the matrix elements of $\mathbf{G}_0$ are closed.

Elastic scattering observables follow from the on-shell properties ($k_1 = k_1' = k$) of the scattering matrices. For the elastic scattering of neutrons (spin $\frac{1}{2}$) from spin zero targets $c = c' = 1$, and $S_{11} \equiv S_\ell^J = S_\ell^{(\pm)}$ are

$$S_{11} = 1 - i\pi \frac{2\mu k}{\hbar^2} \sum_{nn'=1}^{M} \chi_{1n}(k) \frac{1}{\sqrt{\eta_n}} \left[\left(1 - \boldsymbol{\eta}^{-\frac{1}{2}} \mathbf{G}_0 \boldsymbol{\eta}^{-\frac{1}{2}} \right)^{-1} \right]_{nn'} \frac{1}{\sqrt{\eta_{n'}}} \chi_{1n'}(k).$$

Diagonalization of the complex-symmetric matrix,

$$\sum_{n'=1}^{N} \eta_n^{-\frac{1}{2}} \left[\mathbf{G}_0 \right]_{nn'} \eta_{n'}^{-\frac{1}{2}} \tilde{Q}_{n'r} = \zeta_r \tilde{Q}_{nr},$$

establishes the evolution of the complex eigenvalues ζ_r with respect to energy. Resonant behaviour occurs when an eigenvalue ζ_r crosses the unit circle in the Gauss plane. The energy at which such occurs is the resonance centroid and the imaginary part of the associated eigenvalue ζ_r defines the resonance width. It is evident that the elastic channel S-matrix has a pole structure at the corresponding energy where one of these eigenvalues approach unity, since

$$\left[\left(1 - \boldsymbol{\eta}^{-\frac{1}{2}} \mathbf{G}_0 \boldsymbol{\eta}^{-\frac{1}{2}} \right)^{-1} \right]_{nn'} = \sum_{r=1}^{N} \tilde{Q}_{nr} \frac{1}{1 - \zeta_r} \tilde{Q}_{n'r}.$$

While the required starting matrix of potentials within the MCAS method may be constructed from any model of nuclear structure, to date, only simple collective models have been used to define those potentials with deformation taken to second order. Also, to date, only the first few (3 to 5) low lying states of the target nucleus have been used to specify the coupled-channel problem.

2.3. g-folding optical potentials and the DWA

Differential cross sections for both elastic and inelastic scattering from the Carbon isotopes have been predicted using the microscopic g-folding model of the Melbourne group [2]. That model begins with the NN g-matrices for the interaction of a nucleon with infinite nuclear matter. Starting with the Bonn-B free NN interaction [3], those g-matrices are solutions of the Brueckner-Bethe-Goldstone equations in infinite nuclear matter, $viz.$

$$g\left(\mathbf{q}',\mathbf{q};K\right) = V\left(\mathbf{q}',\mathbf{q}\right) + \int V\left(\mathbf{q}',\mathbf{k}'\right) \frac{Q\left(\mathbf{k}',\mathbf{K};k_f\right)}{\left[E\left(\mathbf{k},\mathbf{K}\right) - E\left(\mathbf{k}',\mathbf{K}\right)\right]} \, g\left(\mathbf{k}',\mathbf{q};K\right) \, d\mathbf{k}'.$$

Both Pauli blocking of states and an average background mean field in which the nucleons move are incorporated, leading to g-matrices that are complex, energy-dependent, medium (density) dependent, and nonlocal in that the solutions for different partial waves reflect tensorial character. Such g-matrices can be, and have been, used directly in momentum space evaluations of NA (elastic) scattering [4], but I prefer to analyze data using a coordinate space representation. For this, and to make use of the program suite DWBA98 [5], the g-matrices must be mapped, via a double Bessel transform to equivalent forms in coordinate space. Folding those effective interactions, $g_{\text{eff}}(0,1)$ with the density-matrices of the target then yields a complex, nonlocal, density-dependent, nucleon-nucleus (NA) optical potential from which the elastic scattering observables are obtained. Full details of this prescription can be found in the review article [2].

Inelastic nucleon scattering is calculated within the Distorted-Wave-Approximation (DWA) using the effective coordinate space g-matrices ($g_{\text{eff}}(0,1)$) as the transition operator. Again full details are to be found in the review [2]. The transition amplitude is given by

$$T_{J_f J_i}^{M_f M_i \nu' \nu}(\theta) = \left\langle \chi_{\nu'}^{(-)} \right| \left\langle \Psi_{J_f M_f} \right| A g_{\text{eff}} \mathcal{A}_{01} \left\{ \left| \chi_\nu^{(+)} \right\rangle \left| \Psi_{J_i M_i} \right\rangle \right\},$$

where $\chi^{(\pm)}$ are distorted wave functions for an incident and an emergent nucleon respectively. Those wave functions are generated from g-folding optical potentials. Coordinates 0 and 1 are those of the projectile and a chosen struck bound state nucleon, respectively. By using a co-factor expansion of the target wave function one obtains as a function of the scattering angle

$$T_{J_f J_i}^{M_f M_i \nu' \nu} = \sum_{\alpha_k, m_i} \sum_{JM} \frac{(-1)^{j_1 - m_1}}{\sqrt{2J_f + 1}} \langle j_2 \, m_2 \, j_1 \, -m_1 | \, J_f \, M_f \rangle \, \langle J_i \, M_i \, J \, M \, | \, J_f \, M_f \rangle$$

$$\times \left\langle J_f \left\| \left[a_{\alpha_2}^\dagger \times \tilde{a}_{\alpha_1} \right]^J \right\| J_i \right\rangle \, \left\langle \chi_{\nu'}^{(-)} \left| \langle \varphi_{\alpha_2} | \, A g_{\text{eff}} \, A_{01} \left\{ \left| \chi_\nu^{(+)}(0) \right\rangle | \varphi_{\alpha_1}(1) \rangle \right\} \right.$$

for an angular momentum transfer J. Here α denotes the set of single-particle quantum numbers $\{n, l, j, m_\tau\}$, where τ is the nucleon isospin. Thus the scattering amplitudes are weighted sums of two-nucleon amplitudes; the weights being transition one-body density matrix elements (OB-DME),

$$S_{\alpha_1 \alpha_2 I}^{J_i J_f} = \left\langle J_f \left\| \left[a_{\alpha_2}^\dagger \times \tilde{a}_{\alpha_1} \right]^J \right\| J_i \right\rangle \, ,$$

and which are to be defined from whatever model of nuclear structure is used. With the g-folding potentials defining the distorted waves, and the g_{eff} being the transition operator, in the DWA, the problem reduces to one of specifying the structure of the target. For this study two different models have been used. The first is the Skyrme-Hartree-Fock (SHF) model of Brown [6], With that model and using the SKX interaction, ground state densities and single nucleon wave functions have been defined. However, as the SHF model cannot provide information on the spectrum of the target, a $(0+2)\hbar\omega$ shell model (SM) has been used to find the transition densities that are input to DWA calculations of inelastic scattering. Note that for ^{10,12,14}C, this SM is complete while for ^{16}C and ^{18}C the space is truncated, excluding the $0g1d2s$ shell required for a complete evaluation of 1p-1h, $2\hbar\omega$ excitations from the $1s0d$ shell. Calculations have been made using the OXBASH shell model program [7] with the WBP interaction of Warburton and Brown [8] suitably corrected for center-of-mass effects. With that model the ground state wave functions for ^{10}C, ^{12}C and ^{14}C are

$$|^{10}\text{C}\rangle = 92.4\% \, |0\hbar\omega\rangle + 7.6\% \, |2\hbar\omega\rangle$$
$$|^{12}\text{C}\rangle = 87.0\% \, |0\hbar\omega\rangle + 13.0\% \, |2\hbar\omega\rangle$$
$$|^{14}\text{C}\rangle = 84.9\% \, |0\hbar\omega\rangle + 15.1\% \, |2\hbar\omega\rangle \, ,$$

while the states in ^{16}C and ^{18}C are purely $2\hbar\omega$ in character. These shell models give excitation energies, and the $B(E2)$ values for excitation, of the 2^+ states in the even mass C isotopes that, in general, are in agreement with data. Of note is that the energy of the 2_1^+ state in ^{14}C is much larger than that in the other isotopes.

2.4. Studies of the isotopes of Carbon using nucleon-Carbon interactions

There has been much speculation concerning the possible melting or chang-
ing of the shell structure of nuclei as one moves away from the valley of stability. With light mass nuclei there are indications of possible changes in the
magic numbers. Around ^{32}Mg one finds an 'island of inversion' [9]. In structure calculations, such variations are influenced by three-body forces [10],
as well as by changes in the monopole term of the Hamiltonian. The latter has been shown to cause changes to the single particle energies as one
approaches the drip lines. So, in theory, there are reasons to expect new
magic numbers in nuclei off of the stability line [11].

Were there to be such an extreme change in the shell structure in nuclei
as that system becomes neutron or proton rich away from stability, means
by which to identify that are required. For even mass nuclei, one way is to
consider the energy of the first 2^+ state systematically about a suspected
closed shell [12]. Besides expected signatures in the spectrum, the cross
sections from inelastic scattering of the (radioactive) isotopes or isotones,
as radioactive ion beams (RIBs), around the closed shell nucleus, from
hydrogen targets might reveal a distinct variation with mass.

The even mass Carbon isotopes provide a set of nuclei with which one
may observe a trend from the proton to the neutron drip lines, ($^{12-18}$C).
They also span a known neutron shell closure with ^{14}C as evident from their
low lying spectra. In the spectra of ^{10,12,16,18}C the first excited state has
spin-parity 2^+ and excitation energies of 3.35, 4.43, 1.77, and 1.62 MeV
respectively. In contrast, in the spectrum of ^{14}C, the first excited state,
with spin-parity 1^- is associated with a cluster of states in the range 6 to 7
MeV. Spin-parities of those states are in sequence $1^-, 0^+, 3^-, 0^-$ and then
2^+. Thus these nuclei are an excellent set with which to show any signature
of shell closure with scattering data. Of course there are other properties
one can consider to note shell closures, for example the $B(E2)$ value and
electron scattering form factor [13].

Herein, scattering data is considered as a means by which shell closures
and crossing may be identified. In particular, proton scattering at intermediate energies (E) which, by inverse kinematics, equates to RIB scattering
(E per nucleon) from hydrogen targets, is considered. At these energies,
the nucleon-nucleon (NN) potential is dominated by the V_{pn} component
interaction [2] and so proton scattering primarily, though not exclusively,
probes the neutron density in a nucleus. By symmetry, neutron scattering

primarily is a probe of the proton distribution in the nucleus.

As noted, the neutron shell closure in the Carbon isotopes occurs at ^{14}C, with closed neutron $0p$-shells. There is closure of the $0p_{\frac{3}{2}}$ shell in ^{12}C, but that is not a purely closed sub-shell as there are significant 2p-2h terms in the wave function. I consider inelastic scattering leading to the first 2^+ state in each isotope to see if there is a significant change in the cross section shape at the shell closure consistent with the change in the spectrum. Changes are expected to occur in both the elastic and the inelastic scattering around the ^{14}C results. However, to establish such requires a scattering theory that is predictive, i.e. not subject to parameter adjustment.

2.4.1. *Using MCAS and the spectra of odd-mass isotopes*

The first application of the MCAS method was to find the spectrum of ^{13}C from calculations of the $n+^{12}$C system. Details have been published [1]. I simply note here that, treating the problem as one of three target state couplings, using a collective rotor model specification of the interaction matrix of potentials, accounting for the Pauli blocking of the occupied nucleon orbits in the target, and using the resonance finding methodology, gave a very credible spectrum for ^{13}C with one to one matching of all known states to ~ 10 MeV excitation. Subsequently, with the same model suitably adjusted for extra Pauli blocking, the spectrum of ^{15}C was determined from an MCAS treatment of the $n+^{14}$C system [14]. The result is shown on the left of Fig. 2.1. The energy scale is that for a neutron on ^{14}C and the states of ^{14}C shown were used in the MCAS evaluations. An interaction matrix was found that gave the spectrum of ^{15}C labelled by 'MCAS' which is compared with the known spectrum ('EXP.'). The states identified by twice their spin and their parity match very well especially given that the calculated width of the $\frac{3}{2}^+$ resonance indicated by the dashed lines is quite large.

The nuclear interaction determined from finding the best spectrum of ^{15}C then was used to study the $p+^{14}$O system given that such is the mirror isospin of $n+^{14}$C. Only a Coulomb interaction was added. The resulting spectrum made ^{15}F particle unstable, as it is known to be, with a $\frac{1}{2}^+$ resonant ground state. Low energy scattering experiments of ^{14}O ions from hydrogen targets gave cross sections as shown in the right side of Fig. 2.1. The cross section displayed by the solid curve is that obtained from the MCAS evaluation. It is of note that the theory also expects resonance

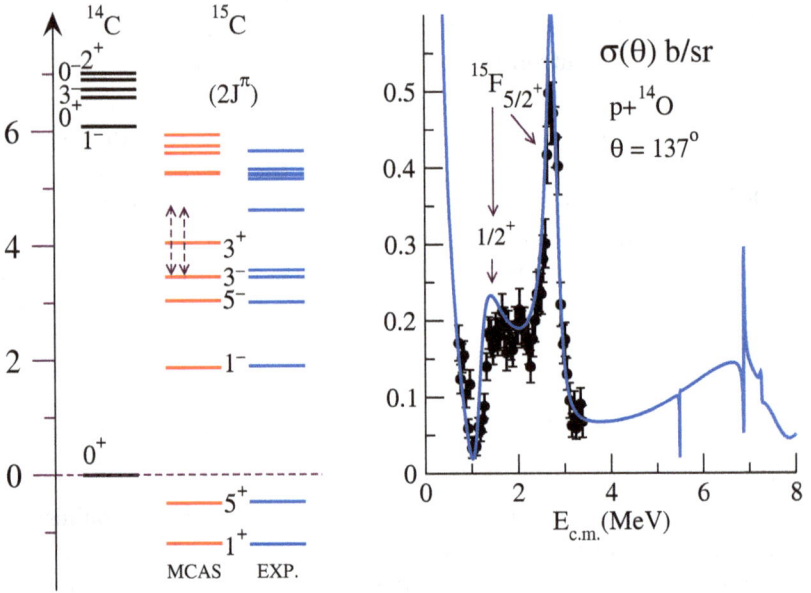

Fig. 2.1. The low excitation spectrum of ^{15}C relative to the the $n+^{14}$C threshold and a cross section from ^{14}O ions scattering from hydrogen.

behaviour at slightly higher incident energies than measured as yet.

In Fig. 2.2 the currently known spectrum [15] (just three bound and one resonance) of ^{19}C are compared with the low excitation spectra for this nucleus determined from a SM calculation and with that found from the coupled-channel solutions (MCAS) based upon a two-state, collective, model for the $n+^{18}$C system. A collective rotor prescription in a three state, 0_1^+, 2_1^+, and 2_2^+, coupled-channel description was used. In this case prolate deformation was required to get the three closely spaced low-lying bound states in ^{19}C. The changes in the spectrum found by varying the deformation parameter β_2 are displayed in Fig. 2.3. The excitation energies of the states all vary regularly as the deformation increases, either for prolate (positive β_2) or oblate (negative β_2) character. Two states stand out as being dominantly the coupling of a single neutron to the ground state of ^{18}C. They are denoted by the filled circles (lowest set) being the ground state of ^{19}C formed (when $\beta_2 = 0$) with a $1s_{\frac{1}{2}}$–neutron, and by the filled

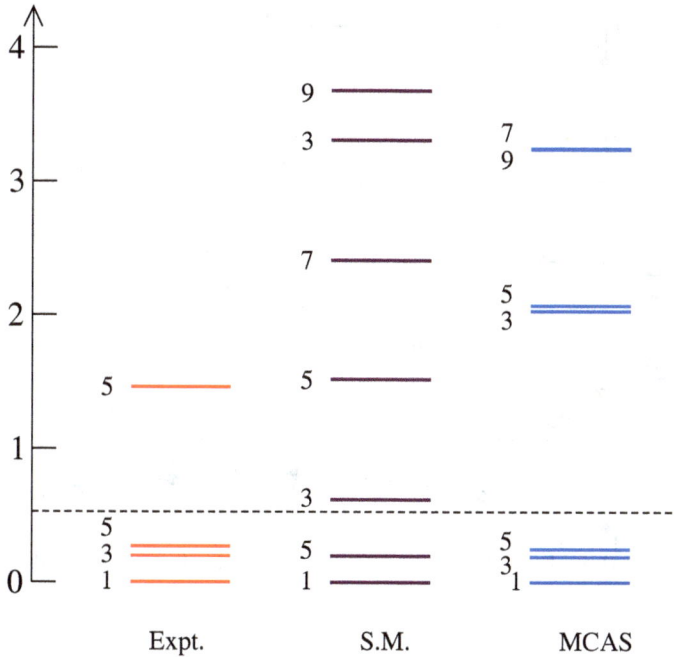

Fig. 2.2.   The low excitation spectrum of ^{19}C. The calculated spectra are identified by S.M. and MCAS with the first found using a $2\hbar\omega$ shell model. The dashed line indicates the $n+^{18}$C threshold, while the numeral identifying each state is $2J$.

triangles being a state formed (when $\beta_2 = 0$) with a $0d_{\frac{5}{2}}^+$−neutron. These states vary with deformation noticeably more slowly than the others. That is so especially for the ground state reflecting that the prime admixing component $\left(\left[0d_{\frac{5}{2}} \otimes 2^+\right]\Big|_{\frac{1}{2}}\right)$ lies above 4 MeV in the unperturbed spectrum. But it is important to note that the deformation coupling mixes all basic states of given spin-parity to form the resultant ones in the spectrum of ^{19}C. From this plot it is also clear that coupled-channel calculations for this system require a strong deformation ~ 0.4 to obtain three states of the appropriate spin-parity lying below the neutron-^{18}C threshold and still retaining a one-neutron separation energy of ~ 0.53 MeV. The spectrum of ^{17}C has been studied in the same way and similar results found to those for ^{19}C. Full details of MCAS studies of both radio-active ions has been published recently [16].

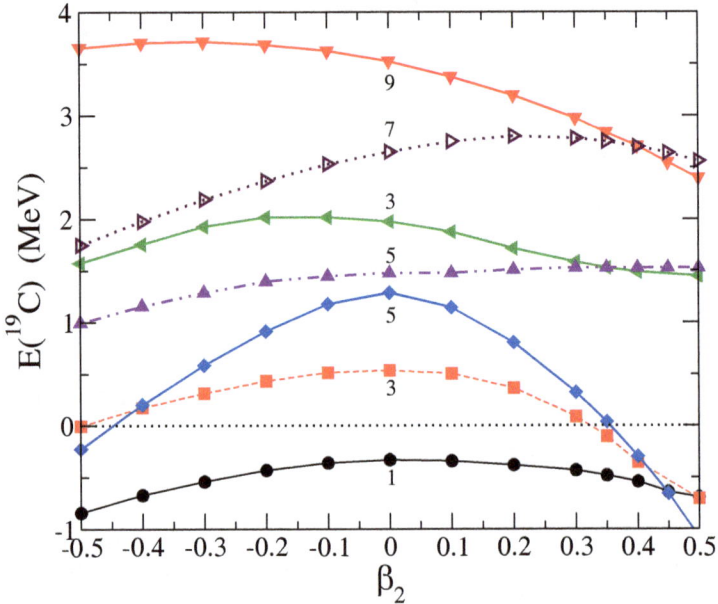

Fig. 2.3.   The variation in the low excitation spectrum of ^{19}C with deformation.

2.4.2. *Using g-folding optical potentials*

The differential cross sections for the elastic scattering of 100 MeV protons from $^{10-18}$C using the shell occupancies and single particle wave functions obtained by using the shell model, are displayed in Fig. 2.4. For the shell model, an oscillator parameter of 1.65 fm was used to specify the single-particle wave functions. There is a steady increase in the cross section with angle indicating an increasing rms radius with mass. Beyond the first diffraction minimum the cross sections from ^{10}C and ^{12}C are significantly reduced compared to the other isotopes. That indicates the lack of $0p$-shell strength in the density, as ^{14,16,18}C all have a closed $0p$ neutron shell.

Consider next, the inelastic scattering to the first 2^+ state in each nucleus. Using the shell model structures to define the transition OBDME, the differential cross sections for the inelastic scattering of 100 MeV protons to the 2^+_1 state in each nucleus are displayed in Fig. 2.5. Here, the differential cross sections from ^{10,12}C and ^{16,18}C while comparable in magnitude are most distinctly different in shape. That indicates the change in the density with the introduction of the sd-shell neutrons in ^{16,18}C. Most

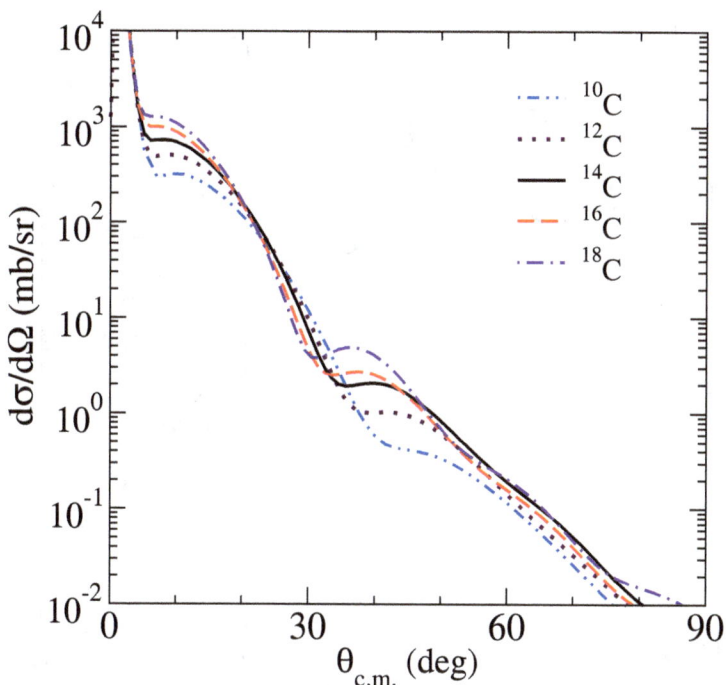

Fig. 2.4.   The differential cross sections for the elastic scattering of 100 MeV protons from the even isotopes of Carbon when their structure was obtained from the shell model.

striking is the reduction in strength of the differential cross section from scattering to the 2^+ state in ^{14}C. This reduction is an order of magnitude at $0°$ decreasing to a factor of 6 at $20°$. As the neutron shell is closed in ^{14}C, there is very little neutron strength in the transition density in this nucleus and so there is a reduction in the inelastic cross section from it, compared to those from excitations of the 2^+ states in the other isotopes.

In a recent paper, Satou *et al.* [15] report data from the inelastic scattering of ^{17,19}C ions from a hydrogen target. Differential cross sections from the scattering of $70A$ MeV ions leading to states at excitation energies of 2.2 and 3.05 MeV in ^{17}C and to the 1.46 MeV excited state in ^{19}C were presented. Our analyses of the data have been published [16], and so for illustration, only the data from the excitation of the 1.46 MeV state in ^{19}C are shown in Fig. 2.6. There are three evaluations with which these data are compared. Those depicted by the solid and long-dashed curves were made assuming that the ground state had a spin-parity of $\frac{1}{2}^+$. Both results are

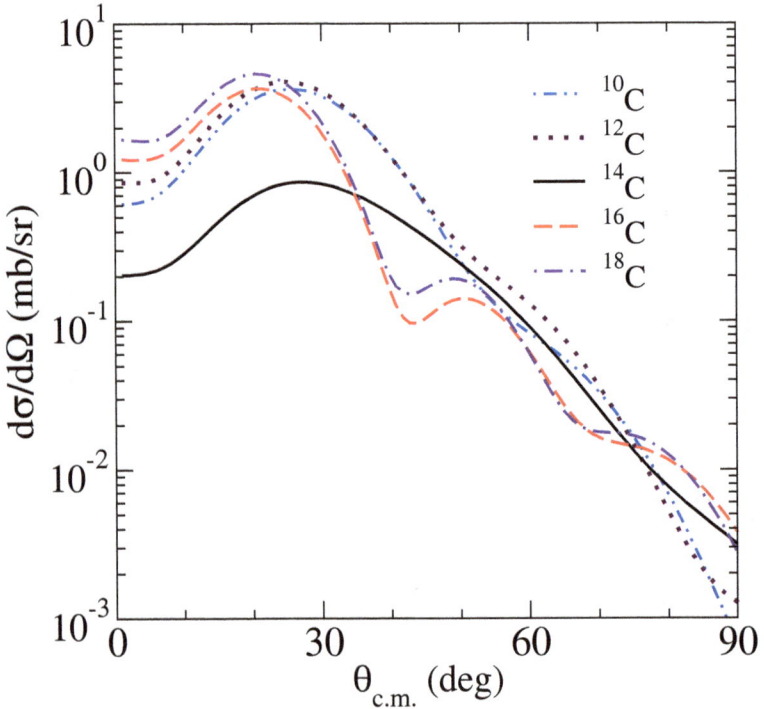

Fig. 2.5.   The differential cross sections for the inelastic scattering of 100 MeV protons to the 2^+_1 states in $^{10-18}$C, obtained using shell model OBDME.

dominated by $L = 2$ angular momentum transfer. The solid (dashed) curve is the result found using oscillator (Woods-Saxon) functions for the single-nucleon bound-state wave functions that reflect a neutron skin (halo-like) property to the density. The result depicted by the dot-dashed curve, was found using the oscillator wave functions but on assuming that the ground state had a spin-parity of $\frac{5}{2}^+$. These results indicate that the ground state of ^{19}C has indeed the $\frac{1}{2}^+$ assignment as has been suggested [15], but that it may have a neutron skin rather than a neutron halo.

2.5. MCAS and hypernuclei

Recently the MCAS scheme has been used to study the spectra of hyper-nuclei. Since the Λ^0 particle is a baryon with spin $\frac{1}{2}$ and a rest mass 1115.6 MeV/c^2 (c/f the neutron rest mass is 940 MeV/c^2), MCAS evaluations of

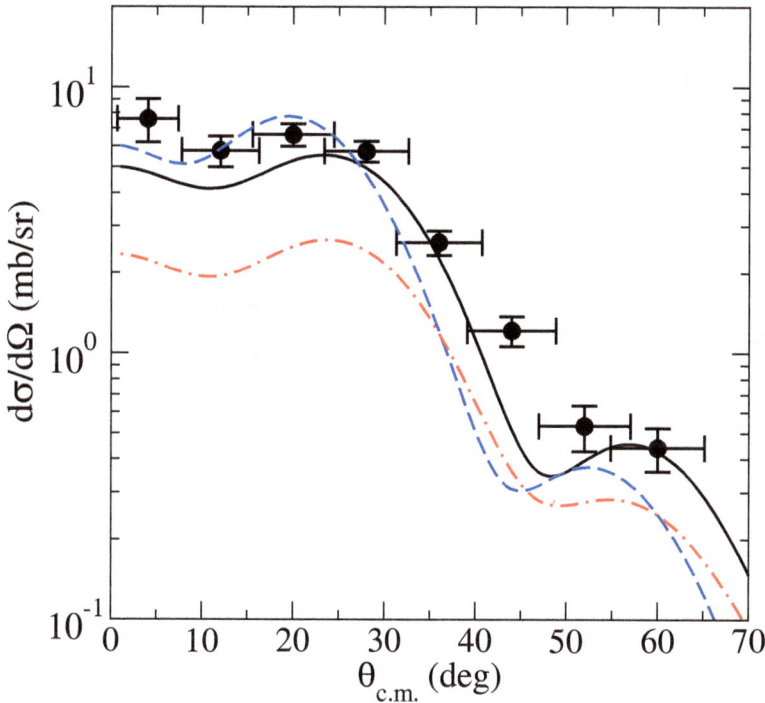

Fig. 2.6. The differential cross sections for the inelastic scattering of 70 MeV protons from ^{19}C leading to the state at 1.46 MeV excitation.

Λ-Nucleus systems is very similar to those of neutron-nucleus systems but without any problem of the Pauli principle blocking states. Hypernuclear systems are of some interest in that fine splittings in the hypernuclear spectra link closely to characteristics in the Λ-nucleon interaction. Also, since the hyperon is not restricted by the Pauli principle in the nuclear medium, it can act as a 'tag' to study systems that have excess neutrons, e.g. ^{48}Ca. With the current MCAS, one can analyse both bound and resonant spectra for (light mass) Λ-hypernuclei, to support and interpret experimental investigations.

To illustrate, using nucleon-nucleus interactions scaled as has been done in past studies, *viz.* the central strengths by $\frac{2}{3}$, the spin-orbit strengths by an order of magnitude, and the radii 15 - 20 % smaller, the spectra of $^{13}_{\Lambda}$C and $^{9}_{\Lambda}$Be were found that are compared with known states of those systems in Fig. 2.7.

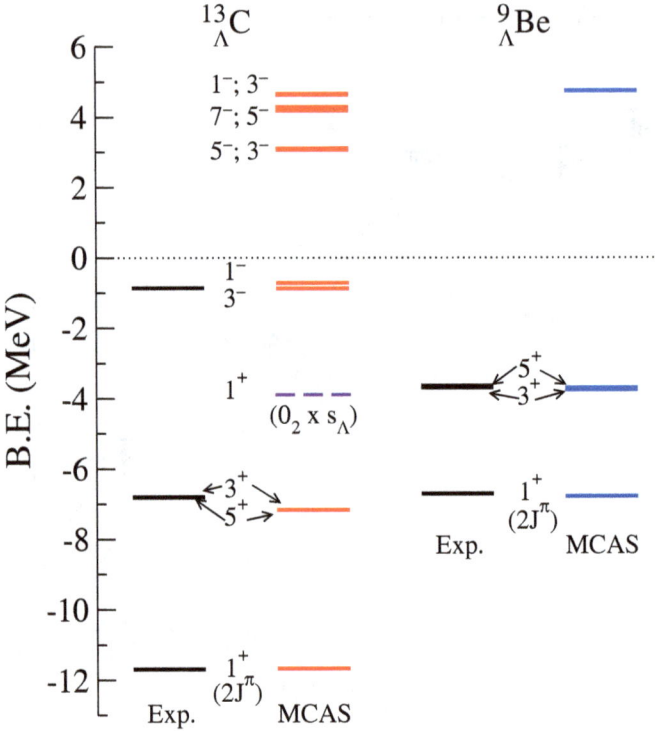

Fig. 2.7.   The spectra of $^{13}_{\Lambda}$C and of $^{9}_{\Lambda}$Be evaluated using MCAS compared with data.

The $\frac{1}{2}^{-}$-$\frac{3}{2}^{-}$ splitting in $^{13}_{\Lambda}$C is dominated by the Λ-core spin-orbit interaction. This splitting is largely insensitive to other factors such as the deformation of the core or other spin-dependent components of the Λ-core potential. On the other hand, the fine splittings of the $\frac{3}{2}^{+}$-$\frac{5}{2}^{+}$ doublet in $^{9}_{\Lambda}$Be and $^{13}_{\Lambda}$C have a different dynamical origin. The coupling of the $s_{\frac{1}{2}}$-Λ single-particle motion with the 2^{+} collective excitation of the core. As it involves an s-wave Λ-particle, it is independent of the spin-orbit interaction. While this splitting varies slightly with the deformation of the core, it is most sensitive to the coupling of the $s_{\frac{1}{2}}\Lambda$ single-particle motion with the 2^{+} collective excitation of the core through the V_{sI} interaction. The recently measured structure of the fine ($\frac{3}{2}^{+}$-$\frac{5}{2}^{+}$) splitting in $^{9}_{\Lambda}$Be determines the sign and strength of that spin-spin interaction.

2.6. Conclusions

The spectra of light mass nuclei show bound and resonant states that are distinct. Likewise low energy cross sections from the collision of nucleons with a (light mass) nucleus shows sharp, as well as broad, resonances lying upon a smooth, energy-dependent background. Those resonances correlate to states in the discrete spectrum of the target. To interpret spectra and such low energy scattering data, the MCAS approach has proved a most effective means to find solutions of the coupled Lippmann-Schwinger equations that define the problems. With isospin symmetry assumed, the method gives spectra for nuclei that are at or beyond the proton drip line. A new application of MCAS, to specify the spectra of Λ hypernuclei was shown as well.

For higher energy nucleon-nucleus scattering data, the g-folding and DWA approaches are pertinent. Using them, it has been shown that cross sections from inelastic scattering may be used to identify shell closure. For the nucleus in which that shell closure occurs, the transition density can be markedly reduced from similar transition cross sections in neighbouring isotopes. Beyond the shell closure the change in the density effected by the introduction of higher-order shells is also significant. These two points together are indicative of the major shell closure. While the transition may be investigated with zero-momentum transfer observables, such as $B(El)$ values, one may only ascertain the change in the density with data from scattering experiments that probe the density at finite momentum transfer values. But experimental investigation of inelastic scattering must also take elastic scattering data to interpret correctly the underlying optical potentials involved [17].

Acknowledgments

The research described in this paper has been made in collaboration with colleagues of the so-called MCAS collaboration, Profs. L. Canton and G. Pisent of the University of Padova, Italy, Prof. J. P. Svenne of the University of Manitoba, Canada, Prof. S. Karataglidis of Rhodes University, South Africa, and by Messrs. P. Fraser and D. van der Knijff of the University of Melbourne. The research was supported by grants from the Australian Academy of Science, from the Italian MURST-PRIN Project "Fisica Teorica del Nucleo e dei Sistemi a Più Corpi", and from the Natural Sciences and Engineering Research Council (NSERC), Canada.

My colleagues in these research studies have on, some or many, occasions, met and talked with Ken Hines. They with me were greatly saddened by his passing. Collectively, we acknowledge his wisdom, his encouragement of our research pursuits, and his friendship.

Vale Ken Hines, *Requiescas in Pace*

References

[1] K. Amos, L. Canton, G. Pisent, J. P. Svenne, and D. van der Knijff, *Nucl. Phys.* **A728**, 65 (2003).

[2] K. Amos, P. J. Dortmans, H. V. von Geramb, S. Karataglidis, and J. Raynal, *Adv. in Nucl. Phys.* **25**, 275 (2000).

[3] R. Machleidt, K. Holinde, and Ch. Elster, *Phys. Rep.* **149** 1, (1987).

[4] H. F. Arellano, F. A. Brieva, M. Sander, and H. V. von Geramb, *Phys. Rev. C* **54**, 2570 (1996).

[5] J. Raynal, *computer program* DWBA98, NEA 1209/05 (1998).

[6] B. A. Brown, *Phys. Rev. C* **58**, 220 (1998).

[7] OXBASH-MSU *(the Oxford-Buenos-Aries-Michigan State University shell model code).* A. Etchegoyen, W.D.M. Rae, and N.S. Godwin (MSU version by B.A. Brown, 1986); B.A. Brown, A. Etchegoyen, and W.D.M. Rae, MSUCL Report Number 524 (1986).

[8] E. K. Warburton and B. A. Brown, *Phys. Rev. C* **46**, 923 (1992).

[9] E. K. Warburton, J. A. Becker, and B. A. Brown, *Phys. Rev. C* **41**, 1147 (1990).

[10] S. C. Pieper, K. Varga, and R. B. Wiringa, *Phys. Rev. C* **66**, 044310 (2002).

[11] T. Otsuka, R. Fujimoto, Y. Utsuno, B. A. Brown, M. Honma, and T. Mizusaki, *Phys. Rev. Lett.* **87**, 082502 (2001).

[12] J. I. Prisciandaro *et al. Phys. Lett.* **B510**, 17 (2001).

[13] S. Karataglidis, B. A. Brown, K. Amos, and P. J. Dortmans, *Phys. Rev. C* **55**, 2826 (1997).

[14] L. Canton, G. Pisent, J. P. Svenne, D. van der Knijff, K. Amos, and S. Karataglidis, *Phys. Rev. Lett.* **94**, 122503 (2005).

[15] Y. Satou *et al. Phys. Letts.* **B660**, 320 (2008).

[16] S. Karataglidis, K. Amos, P. Fraser, L. Canton, and J. P. Svenne, *Nucl. Phys.* **A813**, 235 (2008).

[17] K. Amos, W. A. Richter, S. Karataglidis, and B. A. Brown, *Phys. Rev. Lett.* **96**, 032503 (2006).

Chapter 3

Resonant X-Ray Scattering and X-Ray Absorption: Closing the Circle?

Zwi Barnea*, Chris T. Chantler, Martin D. de Jonge, Andrew W. Stevenson[†], and Chanh Q. Tran

The University of Melbourne, School of Physics, Parkville 3010, Vic. Australia
*E-mail: barnea@physics.unimelb.edu.au

Bragg intensities from zinc selenide, cadmium selenide, cadmium sulphide and gallium arsenide crystals illustrate the excellent internal consistency of x-ray Bragg intensities measured with extended-face crystals. Applying this technique to measurements of the Bijvoet ratios of zinc selenide as a function of x-ray wavelength between 0.6 Å and 1.6 Å, a region which includes the absorption edges of Zn and Se, showed that the measured and calculated Bijvoet ratios exhibited systematic discrepancies which at the time were difficult to interpret.

Starting in the 1990s, techniques were developed for the accurate measurement of x-ray attenuation coefficients which allowed careful testing, avoidance or correction of systematic errors. These measurements revealed systematic discrepancies between the measured and calculated attenuation coefficients on both sides of the absorption edges of a number of elements (Cu, Ag, Mo, and Sn).

Because the imaginary dispersion correction is directly related to the photoelectric absorption coefficient, the possibility exists that the observed discrepancies in the zinc selenide Bijvoet ratios and possible discrepancies between the calculated and measured absorption coefficients of zinc and selenium in the region of their absorption edges are manifestations of the same effect. If confirmed, the result would be a consistent explanation of both discrepancies and the theoretical calculations of absorption coefficients would then require critical review.

We propose a relatively simple synchrotron-based combined experiment in which the Bijvoet ratios and the attenuation coefficients are both measured at identical x-ray energies.

[†]CSIRO Manufacturing and Infrastructure Technology, Clayton South, Vic. Australia.

3.1. Bragg intensity measurements from extended-face crystals

The use of extended-face crystals for measuring the intensities of x-ray Bragg reflections was initially forced upon us by the need to measure x-ray intensities from a unipolar flat tetragonal barium titanate crystal. Any attempt at mechanical shaping of this crystal would have destroyed its unipolar nature. Our first set of Bragg intensity measurements was obtained from the sets of planes parallel to the growth facet of the crystal. When the need arose to extend the measurements to intensities diffracted by other planes, not parallel to this crystal facet, we initially chose to work in an asymmetric configuration and to apply the appropriate absorption correction. Later, we found it preferable to measure intensities in two complementary asymmetric positions and average the intensities. Analysis of the measurements obtained with this technique showed quite remarkable agreement of the intensities of equivalent reflections and led, in combination with neutron data, to the determination of the structure of tetragonal barium titanate [1], a structure which had for almost two decades resisted elucidation.

Further analysis of the technique most suitable for Bragg intensity measurements from extended-face crystals revealed the possibility of measuring essentially all reflections, including those from planes not parallel to the crystal face, in symmetric mode, provided the crystal is mounted with its face normal parallel to the φ-axis of a four-circle diffractometer with the φ and χ axes used to bring the diffraction vector of the desired reflection into the diffraction plane of the diffractometer. Later developments involving the avoidance of multiple reflection conditions forced us to go back to averaging intensities measured in two slightly asymmetric complementary positions.

This extended-face crystal measurement technique was then used in studies of cadmium sulphide [2], cadmium selenide [3], zinc selenide [4, 5] and gallium arsenide [6] (as well as indium arsenide [7]). The observed Bragg intensities were interpreted within a model that included resonant scattering dispersion corrections of the atomic form factors, anharmonic thermal vibrations and extinction. Intensities of equivalent reflections were found to agree to better than one percent in almost 80 percent of the data, and exceeded one percent primarily in reflections for which counting statistics were important.

The general conclusions about the extended-face crystal technique were that it made it possible to measure Bragg intensities with a precision hitherto unachievable even with the most regular cylindrically or spherically shaped crystals. Because the entire incident x-ray beam was intercepted by the crystal, even weak reflections could be measured with adequate statistics. The technique was especially valuable in the case of highly absorbing crystals for which the correction for absorption by small, even regularly shaped, crystals presents special, and often overwhelming, difficulties. The extended-face technique was easy to apply, requiring almost no changes to normal diffractometer instrumentation and software. Later it was also found that the technique required no further modification when used in a synchrotron environment.

Most importantly, it was shown by Grigg and Barnea [8] that the extended-face crystal technique is also particularly well suited for absolute intensity measurements and thus holds great promise for measurements of atomic form factors, scale-factor independent studies of extinction [9], absolute determination of structure factors, as well as a host of other hitherto difficult to perform absolute measurements. It is intriguing that the main source of uncertainty in the absolute intensity measurements is now actually due to uncertainties in the absorption coefficients of the elements.

3.2. Bijvoet ratios

It will be noted that the above-mentioned crystals are all non-centrosymmetric and therefore exhibit the Bijvoet effect [10] which causes hkl and -h-k-l, and equivalent reflections belonging to non-centrosymmetric zones, to have different Bragg intensities I_+ and I_- . This effect is due to the presence of resonant scattering (often referred to in the literature as 'anomalous scattering') which results in the atomic form factor being complex. The effect depends on the wavelength of the incident x-rays, and is particularly marked in the vicinity of the absorption edges of the relevant atoms.

The Bijvoet ratio is the intensity difference divided by the average intensity of reflections with positive and negative hkl values (and their equivalents), whose intensities differ by virtue of the resonant scattering effect.

When the Bijvoet ratio is expressed in terms of the structure factors of these two types of reflections with sums of indices of $4n \pm 1$ (where n is an integer), it is given for the case of the particularly simple cubic zinc-blende structure by

$$B = \frac{I_+ - I_-}{(I_+ + I_-)/2} = \frac{4(f_a' f_b'' - f_b' f_a'')}{f_a'^2 + f_b'^2 + f_a''^2 + f_b''^2} \qquad (3.1)$$

where f_a' is the atomic form factor of atom a plus the real component
of the dispersion correction and f_a'' is the imaginary component of the
dispersion correction of atom a. All atomic form factors are assumed to
be multiplied by the appropriate Debye-Waller factors. It is seen from
Eq. (3.1) that the Bijvoet ratio is particularly sensitive to the imaginary
component of the dispersion corrections and relatively insensitive to the real
component of the dispersion corrections, except in the immediate vicinity
of the absorption edge (see Sec. 3.5 below). Similarly, the Bijvoet ratio
is relatively insensitive to the effects of extinction which can be allowed
for [2, 3, 6]. In order to render it more transparent, Eq. (3.1) omits the
effects of anharmonicity and bonding, although these can be included in
the actual interpretation of the Bijvoet ratios.

Although the effects of resonant scattering had been carefully observed
in all the above crystals, Stevenson and Barnea used the extended-face
crystal technique for measuring Bijvoet ratios [11] in order to subject the
resonant scattering in zinc selenide [4, 5, 7] to a particularly detailed test.
The motivation for this was that, unlike cadmium selenide and cadmium
sulphide - hexagonal crystals with the wurtzite structure - zinc selenide has
the particularly simple cubic zinc-blende structure in which the positions
of all the atoms are symmetry determined and thus independent of any po-
sitional parameter. Accordingly, these authors measured an extensive set
of Bragg reflections and used it to determine the Debye-Waller factors and
the anharmonic parameters of zinc selenide, as well as the extinction of the
specimen crystal. They then proceeded to measure the Bijvoet ratios of
selected non-centrosymmetric zone reflections as a function of x-ray wavelength using monochromatised laboratory-tube produced Bremsstrahlung
between about 0.6 Å and 1.6 Å, the region which includes the absorption
edges of both zinc and selenium. A plot of the 311/31-1 Bijvoet ratio as
a function of wavelength is given in Fig 3.1. Therein, the dashed vertical
lines represent the K-absorption edges of Se and Zn [7]. The 31-1 reflection
is equivalent to the -3-1-1 reflection. It is seen that there are systematic
differences between the observed Bijvoet ratios and those calculated with
the theoretically determined dispersion corrections of Cromer and Liberman [12].

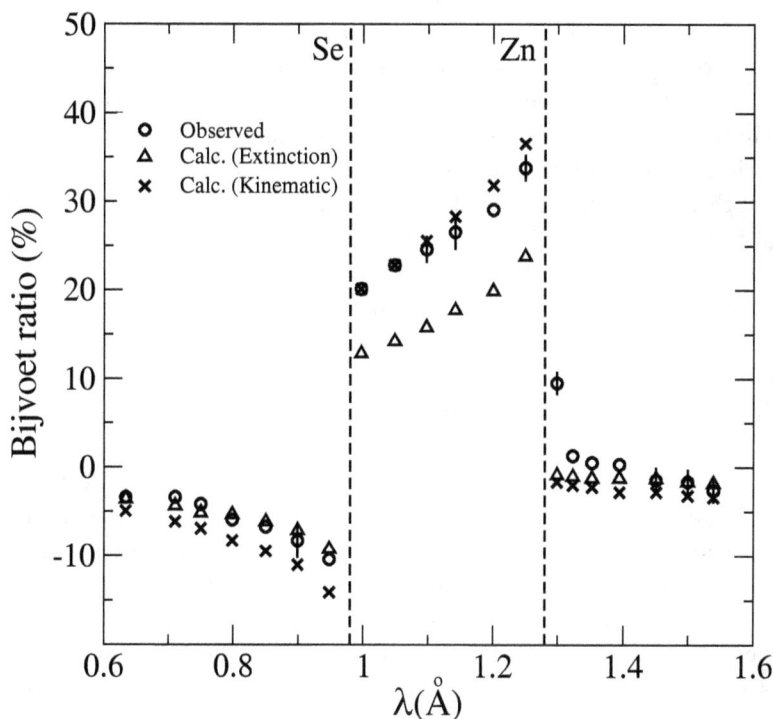

Fig. 3.1.   The variation of the 311/31-1 Bijvoet ratio with wavelength.

3.3. X-ray mass attenuation measurements

In the 1990s two of us (CTC and ZB) commenced an experimental study of x-ray mass attenuation coefficients sufficiently accurate to enable us to compare them with the various available theoretical tabulations. Our hope was that such a comparison would enable us to draw conclusions regarding the best theoretical approach to such calculations and possibly also inform us about the relative merits of the underlying atomic form factor calculations. Our starting points were the tabulation of theoretically calculated atomic form factors, attenuation and scattering of Chantler [13, 14] and the experimental measurements of the x-ray attenuation of silicon of Mika, Martin and Barnea [15], as well as the extensive body of x-ray attenuation studies reported in the literature.

Improving the accuracy of the attenuation measurements depended upon the solution of a number of problems connected with an accurate characterisation of the x-ray beam, the accurate determination of the thickness of the absorber at the actual point where the beam passed through the specimen (the local thickness), and the determination of detector stability, linearity, and counting statistics.

Of these problems, the most crucial - the accurate determination of the local thickness - was solved by combining the macroscopic measurements of the area and mass of the specimen with a detailed x-ray mapping of its thickness variation. All of these measurements can be made very accurately, and the 'full-foil' mapping technique allowed us to obtain very accurate results [16, 17].

The monochromatised synchrotron x-ray beam (all our recent attenuation measurements have been conducted at synchrotrons) requires careful determination of the x-ray energy, as well as tests to establish the presence and effect of higher-order harmonics diffracted by the monochromator crystals and not fully excluded by detuning. These tests, based on a development of Barnea and Mohyla [18], make use of the fact that, as a beam is absorbed, the relative intensity of the hard x-ray component increases and can thus be observed as a non-linearity of the attenuation with thickness[19]. A somewhat similar non-linear absorption effect, confined however to the immediate vicinity of the absorption edge, has been shown to be due to the bandwidth $\Delta E/E$ of the x-ray beam [20].

The energy of the x-ray beam has been carefully calibrated using single-crystal or powder-specimen standards. Detector stability and statistics issues have been discussed in Chantler et al. [21, 22].

Apertures of different sizes placed between the specimen and the counter provided information about the secondary radiation of elastically and inelastically scattered photons reaching the counter during the absorption experiment. The results of such absolute measurements obtained with different thicknesses of silver foils have been compared with calculations based on the expected x-ray fluorescence of the specimens and of the intervening air, as well as on calculations of the Rayleigh and Compton scattering. The effect of secondary radiation on the absorption measurement could thus be reliably estimated and shown to agree with experimental results [23].

The entire suite of techniques briefly summarised above and the use of multiple measurements over extended ranges of each parameter probed in the attenuation experiments has come to be referred to as the x-ray extended-range technique (XERT).

3.4. Attenuation measurements in the vicinity of the absorption edge

The XERT has so far been applied to measurements of mass attenuation coefficients of copper[24], silicon[25], silver[26], and molybdenum[27]. Such data are compared in Fig. 3.2 with theoretical results of Chantler[13, 14] (zero-line), of Scofield[28] used by Berger and Hubbell in XCOM[29] (dotted line) and Creagh *et al.*[30] (dashed line). Also the figure includes values from a semiempirical tabulation of Henke *et al.*[31] (dot-dash line) which is available up to 30 keV. Finally, other experimental data are also included. They are from Machali *et al.*[32] (asterisks), Puttaswamy *et al.*[33] (diamonds), Sandiago *et al.*[34] (triangles) and Visweswara *et al.*[35] (squares). The results are plotted as percentage discrepancy based upon the theoretical results of Chantler[13, 14],

$$\%\text{discrepancy} = \%\frac{[\mu/\rho]_{tot,i} - [\mu/\rho]_{tot,Chantler}}{[\mu/\rho]_{tot,Chantler}}. \tag{3.2}$$

As can be seen from that figure, an appreciable and systematic discrepancy between the measured and calculated mass attenuation coefficients has been observed on both sides of the absorption edge of silver. Similar discrepancies have been observed for molybdenum and copper, and more recently for tin.

In view of the fact that these discrepancies have now been observed in a number of elements, we wondered whether such discrepancies between the calculated and measured values of the mass attenuation coefficients might also be observed in zinc and selenium near their absorption edges.

The imaginary dispersion corrections are related to the photoelectric absorption coefficient by the relationship[13]

$$f'' = \frac{E\sigma_{pe}}{2hcr_e} \tag{3.3}$$

where f'' is the imaginary part of the atomic scattering factor, σ_{pe} is the photoelectric absorption cross-section, E is the photon energy, h and c are Planck's constant and the speed of light, and r_e is the classical electron radius. If the observed mass attenuation coefficients of zinc and selenium are in disagreement with the tabulated values, then the imaginary part of the form factor must also be in disagreement with their tabulated values. Such a disagreement could well account for the discrepancies between the observed and calculated Bijvoet ratios in Fig. 3.1.

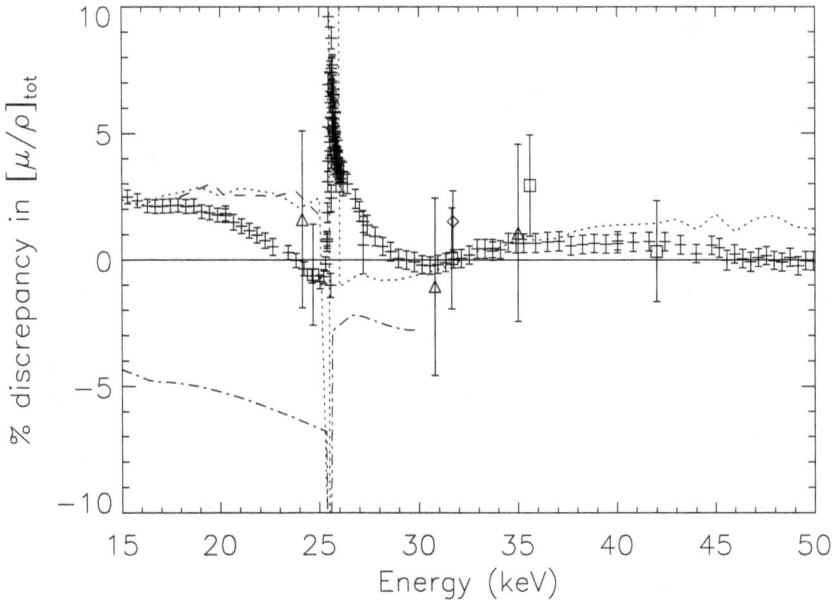

Fig. 3.2. Comparison between the discrepancy as defined in Eq. (3.2), of silver and with other theoretical results, semiempirical results, and other experimental data.

Accordingly in December 2004 and in January 2006 we carried out an experiment at the ANBF in Tsukuba, Japan, measuring the mass attenuation coefficients of zinc and selenium between and around the absorption edges of these elements. These measurements were not carried out at the same energies as the original measurements of the zinc selenide Bijvoet ratios. Nevertheless, we hope that these results, which are currently being interpreted by Nick Rae, will give us an indication whether the absorption edge anomalies are also observed in these elements and whether they lead to results consistent with the observed Bijvoet ratios.

3.5. A combined experiment

A very good way to test the consistency of the measured Bijvoet ratios with the mass attenuation coefficients of zinc and selenium would be to carry out a combined experiment. In this experiment the extended-face crystal of zinc selenide, mounted on a four-circle diffractometer, is used to measure the Bijvoet ratios. The energy of the incident x-rays is set by orienting the zinc selenide crystal to satisfy the diffraction condition for high Bragg angle reflections corresponding to the desired x-ray energy. The monochromator is then adjusted until a maximum intensity is observed at that Bragg angle. Following the Bijvoet ratio measurements, the beam diffracted by the zinc selenide crystal is then in turn used to carry out the mass attenuation XERT measurements with zinc and selenium foils mounted on the detector arm of the diffractometer between the single crystal of zinc selenide and the diffracted-beam detector. This measurement procedure will ensure that the Bijvoet ratios and the mass attenuation coefficients are measured at precisely the same energy. This is of particular importance in the vicinity of the absorption edges where it may well be possible to observe in turn the effect of the real component of the dispersion correction of zinc and selenium on the Bijvoet ratio, especially in the case of high-index reflections.

The measurement of the mass attenuation coefficients and of the Bijvoet ratios at the same x-ray energy is a consistency measurement of related quantities - the mass attenuation coefficient and the imaginary part of the dispersion correction. After the measured mass attenuation coefficients are converted to the corresponding imaginary dispersion corrections, using Eq. (3.2), they are used in the calculation of the expected Bijvoet ratios. If these Bijvoet ratios are in agreement with the measured Bijvoet ratios, then the measurements are consistent. An inconsistency between the measured Bijvoet ratio and the measured mass attenuation coefficient, should such be observed, may of course be much more difficult to interpret.

Acknowledgments

We would like to thank James Hester for his generous assistance with the zinc and selenium measurements at ANBF. This work was supported by the Australian Synchrotron Research Program which is funded by the Commonwealth of Australia under the Major National Research Facilities Program and by a grant of the Australian Research Council. Use of the Advanced Photon Source was supported by the U.S. Department of Energy, Basic Energy Sciences, Office of Energy Research, under Contract No. W-31-109-Eng-38.

In December 2004, during a lunch in the 'Tiamo' Restaurant, the late Ken Hines asked Zwi Barnea: "Tell me, what experiment have you been doing in Japan?". The above paper is an expanded version of my answer. The paper is written in celebration of Ken's wonderful life and of our long and deep friendship.

References

[1] J. Harada, T. Pedersen and Z. Barnea, *Acta Cryst.* **A26**, 336-344 (1970).
[2] A.W. Stevenson, M. Milanko and Z. Barnea, *Acta Cryst.* **B40**, 521-530 (1984).
[3] A.W. Stevenson and Z. Barnea, *Acta Cryst.* **B40**, 530-537 (1984).
[4] A.W. Stevenson and Z. Barnea, *Acta Cryst.* **A39**, 538-547 (1983).
[5] A.W. Stevenson and Z. Barnea, *Acta Cryst.* **A39**, 548-552 (1983).
[6] A.W. Stevenson, *Acta Cryst.* **A50**, 621-632 (1994).
[7] A.W. Stevenson, Ph.D. thesis, The University of Melbourne, 1983.
[8] Mark M. Grigg, Absolute Intensity Measurements on a Modernised Four-circle X-ray Diffractometer, Ph.D. Thesis, The University of Melbourne, 1994.
[9] Anthony J. Richardson, Absolute X-ray Bragg Intensity Measurements as a Means of Determining Extinction, M.Sc. Thesis, The University of Melbourne, 1996.
[10] J.M. Bijvoet, *Nature* **173**, 888 (1954).
[11] S.L. Mair, P.R. Prager and Z. Barnea, *Nature - Phys. Sci.* **234**, 35-36 (1971).
[12] D.T. Cromer and D. Liberman, *J. Chem. Phys.* **53**, 1891-1898 (1970).
[13] C.T. Chantler, *J. Chem. Phys. Ref. Data* **24**, 71-85 (1995).
[14] C.T. Chantler, *J. Phys. Chem. Ref. Data,* **29**, 597 (2000); C.T. Chantler, K. Olsen, R. Dragoset, A. Kishore, S. Kotochigova, and D. Zucker, X-Ray Form Factor, Attenuation and Scattering Tables (version 2.0). Online: http://physics.nist.gov/ffast, National Institute of Standards and Technology, Gaithersburg, MD., 2003.
[15] J. Mika, L. Martin, and Z. Barnea, *J. Phys. C* **18**, 5215-5223 (1985).
[16] C.Q. Tran, C.T. Chantler, Z. Barnea, M. D. de Jonge, *Rev. Sci. Instrum.,* **75**, 2943 (2004)
[17] M. de Jonge, Z. Barnea, C. Chantler, and C. Tran, *Measurement Science and Technology* **15**, 1811-1822 (2004).
[18] Z. Barnea and J. Mohyla, *J. Appl. Cryst.* **7**, 298-299 (1974).
[19] C.Q. Tran, Z. Barnea, M. de Jonge, B.B. Dhal, D. Paterson, D.J. Cookson and C.T. Chantler, *X-Ray Spectroscopy* **32**, 64-74 (2003).
[20] M. de Jonge, Z. Barnea, C.Q. Tran, and C.T. Chantler, *Phys. Rev.* **A69**, art. no. 022717 (2004).
[21] C.T. Chantler, C.Q. Tran, D. Paterson, Z. Barnea, and D.J. Cookson, *X-ray Spectrometry* **29**, 449-458 (2000).
[22] C.T. Chantler, C.Q. Tran, D. Paterson, Z. Barnea, and D.J. Cookson, *X-ray Spectrometry* **29**, 459-466 (2000).
[23] C.Q. Tran, M.D. de Jonge, Z. Barnea and C.T. Chantler, *J. Phys. B.* **37**, 3163-3176 (2004).
[24] C.T. Chantler, C.Q. Tran, Z. Barnea, D. Paterson, D.J. Cookson and D.X. Balaic, *Phys. Rev.* **A64**, art. no. 062506 (2001).
[25] C.Q. Tran, C.T. Chantler and Z. Barnea, D. Paterson, and D.J. Cookson, *Phys. Rev.* **A67**, art. No. 042716 (2003).

[26] C.Q.Tran, C.T. Chantler, Z. Barnea, M.D. de Jonge, B.B. Dhal, C. Chung, D. Paterson, and J. Wang, *J. Phys. B: At. Mol. Opt. Phys.* **38**, 89-107 (2005).

[27] Martin D. de Jonge, Chanh Q. Tran, Christopher T Chantler, Zwi Barnea, Bipin B. Dhal, David J. Cookson, Wah-Keat Lee and Ali Mashayekhi, *Phys. Rev.* **A71**(3), pp.032702/1-16, (2005).

[28] J.J. Scofield, LLNL Report, UCRI-51326, (Lawrence Livermore Laboratory) (1973)

[29] M.J. Berger and J.H. Hubbell, XCOM:Photon Cross Sections Database (NBSIR 87-3597, NIST), (1987); M. J. Berger and J.H. Hubbell "XCOM: Photon Cross Sections Database", version 1.2 (1999) [Online] Available: http://physics.nist.gov/xcom

[30] D.C. Creagh and J.H. Hubbell, in International Table for X-ray Crystallography, edited by Wilson A J C (Kluwer Academic, Dordrecht Vol. C Sect. 4.2.4 (1995)

[31] B.L. Henke, E.M. Gullikson and J.C. Davis, *At. Data Nucl. Data Tables*, **54**, 181 (1993)

[32] F. Machali, G.G. Al-Barakati, A.A. El-Sayed and W.J. Altaf, *J.Phys. F: Met. Phys.*, **17**, 1279 (1987)

[33] K.S. Puttaswamy, R. Gowda and B. Sanjeevaiah, *Can. J. Phys.*, **59**, 22 (1981)

[34] T.K.U Sandiago and R. Gowda, *Pramana*, **48**, 1077 (1997)

[35] V. Visweswara Rao, Shahnawaz and D. Venkateswara Rao, *Physica*, **111c**, 107 (1981)

Chapter 4

The Screened Field of a Test Particle

Robert L. Dewar

Research School of Physics & Engineering
The Australian National University
Canberra ACT 0200, Australia
E-mail: robert.dewar@anu.edu.au

The screened field (forward field and wake) of a test particle moving at constant velocity through an unmagnetised collisionless plasma is calculated analytically and numerically. This paper is based on unpublished material from my MSc thesis, supervised by the late Dr. K. C. Hines.

4.1. Introduction

Interest in the kinetic theory of interacting charged particles at the University of Melbourne developed from the work of Dr. Ken Hines [1] in the 1950s in which he improved on a calculation by Landau [2]. Using a formalism developed by Fano [3], he calculated the slowing down distribution function of a charged 'test particle' passing through a thin layer of material. The plasma theory group developing around Ken attracted a number of research students, including myself. The test particle diffusion problem provided a focus for a reading group on statistical physics [4] and plasma kinetic theory [5, 6], and research on the problem developed in several ways, including a relativistic generalisation [7].

The friction and diffusion coefficients of the Fokker–Planck equation suffer logarithmic divergences arising from the long-range nature of the Coulomb interaction. The fusion plasma theorists of the 1950s handled these divergences by the rather crude device of cutting off the interaction at the Debye length $1/k_D$, the characteristic length occurring in the screened electrostatic potential, $q \exp(-k_D r)/r$, in the neighbourhood of a static particle embedded in an ionised medium. While the logarithmic nature of the divergence makes the coefficients only weakly dependent on the cutoff

[7], the *ad hoc* nature of this procedure was not very satisfactory from a fundamental point of view and this sparked research in the international theoretical physics community to find better approaches.

In 1960, using very sophisticated formalisms, Balescu [8] and Lenard [9] derived kinetic equations of the Landau form in which dynamical screening was incorporated through a frequency (ω) and wavenumber ($\mathbf{k}$) dependent dielectric constant $\epsilon(\omega, \mathbf{k})$. The same Fokker–Planck–Landau equation (which is now known as the Balescu–Lenard equation) was derived by Thompson and Hubbard [10–12] from simpler statistical physics arguments in which the diffusion coefficient was calculated from a fluctuation spectrum obtained by superimposing the dielectrically screened fields of independently moving particles.

This approach was called the 'dressed test particle picture' by Rostoker [13]. In this approach the unperturbed 'test particles' replace the actual particles in the plasma. (In reality the trajectories are perturbed slightly by the fluctuations, giving rise to the linear dielectric response.)

The dressed test particle picture was developed in Fourier, $(\omega, \mathbf{k})$, representation rather than in real space-time, $(\mathbf{x}, t)$, and thus it was difficult to visualise the actual nature of the screened potential surrounding each particle. In Chapters 2 and 3 (unpublished) of my Masters Thesis [14], I calculated the screened potential in real space for a nonrelativistic plasma.[a] This contribution to the K. C. Hines memorial volume is based on those chapters with only a few changes to make it self-contained and to improve readability.

For zero test particle velocity the solution is just the well-known Debye potential $\exp(-k_{\mathrm{D}}r)/r$, but as the velocity is increased we may expect qualitative changes to occur. The work of Pines and Bohm [16] in which they considered forced vibrations of the collective coordinates indicates that for a particle moving slowly with respect to the electron thermal velocity the Debye potential is distorted into a set of spheroidal equipotentials centred on the particle and still decays exponentially with distance. We show that neither result is true, since the criterion $k \lesssim k_{\mathrm{D}}$ is not an adequate criterion for the existence of collective coordinates. Similar results to ours were found in this case by Rand [17, 18] by considering individual particle trajectories in the self consistent field. This is simply a way of circumventing the use of the Vlasov equation but is equivalent to it. Rand makes an analogy between

---

[a]Chapter 1 contained a covariant relativistic plasma response function formalism that was later incorporated into a paper on energy-momentum tensors for dispersive electromagnetic waves [15]

the symmetrical result of Pines and Bohm and the Gibbs paradox of fluid dynamics, and this analogy is upheld by the fact that Majumdar [19, 20] and Cohen [21] get results in agreement with Pines and Bohm by using fluid dynamical treatments.

In this paper we consider a homogeneous isotropic magnetic-field-free plasma. We do not however restrict ourselves to a one-component plasma. In connection with the field around a small satellite, considered by Kraus and Watson [22], it is necessary to consider the ions since the satellite velocity is comparable more with the ion velocity than the electron velocity. In this plasma there are two modes of longitudinal excitation, namely, ion acoustic waves and electron plasma waves, but if the test particle is much slower than the electron thermal velocity then only the ion waves can be excited—the case considered by the above authors. For ion waves not to be Landau damped out of existence the electron Debye length must be much greater than that of the ions, implying the electrons are much hotter than the ions or that the ions have a much greater charge. Kraus and Watson [22] also consider the case of a dense plasma, in which local thermal equilibrium may be assumed.

Despite the preceding remarks about two-component plasmas the case of infinite electron Debye length, which essentially reduces the problem to a one component model, has received the bulk of our attention. Most of our results therefore can directly be compared only with those of Majumdar [19, 20]. We show that his neglect of Landau damping is not justified near the edge of the wake. The supersonic case receives brief attention in Sec. 4.3.3 and 4.3.5.

We note that Pappert [23] considered the effect of an ambient magnetic field on the wake of a test particle, but we do not treat this case. A considerable body of Russian work on the details of the satellite problem had also appeared by the time of this thesis [24–27].

4.2. The formal solution

The electrostatic potential $\varphi(\mathbf{x}, t)$ in the vicinity of a test particle of charge q moving rectilinearly at velocity $\mathbf{v}_0$ ($v_0 \ll c$) through a homogeneous, stable dispersive dielectric medium is obtained from standard linear response theory as

$$\varphi = q \int \frac{d^3 k}{(2\pi)^3} \frac{\exp[i\mathbf{k}\cdot(\mathbf{x} - \mathbf{v}_0 t)]}{k^2 \epsilon(\mathbf{k}\cdot\mathbf{v}_0, \mathbf{k})} \ . \tag{4.1}$$

R. L. Dewar

(Henceforth we consider the time to be $t = 0$, when the test particle is at $\mathbf{x} = 0$, or, equivalently, represent the potential in a frame moving with the test particle.) In Eq. (4.1) $\epsilon(\mathbf{k}\cdot\mathbf{v}_0, \mathbf{k})$ is the dielectric constant, which, for an isotropic, collisionless unmagnetised plasma is given by

$$\epsilon(\omega, \mathbf{k}) \equiv \epsilon(\omega, |\mathbf{k}|) = 1 + \frac{\Phi(\omega/k)}{k^2} \ . \tag{4.2}$$

The function Φ is defined by

$$\Phi(\omega/k) \equiv \sum_s \omega_{ps}^2 \int_{-\infty}^{\infty} dv \frac{g_s'(v)}{\omega/k - v + i0} \ , \tag{4.3}$$

ω_{ps}^2 denoting the plasma frequency, $(e_s^2 n_s/\varepsilon_0 m_s)^{1/2}$ (SI units) for species s, with n_s the unperturbed number density, m_s the mass, and $g_s(v)$ the one-dimensional projection of the unperturbed velocity distribution function $f_s(\mathbf{v})$. That is, in an arbitrary x, y, z Cartesian coordinate system,

$$g_s(v_z) \equiv \frac{1}{n_s} \iint dv_x dv_y f_s(\mathbf{v}) \ , \tag{4.4}$$

the normalisation factor $1/n_s$ being introduced so that $\int dv g_s \equiv 1$. Henceforth we take s to denote electrons and a single species of ion, denoted by subscripts 'e' and 'i' respectively. The assumed form of the 'hodograph' of Φ for each species is sketched in Fig. 4.1, which is such as to ensure stabil-

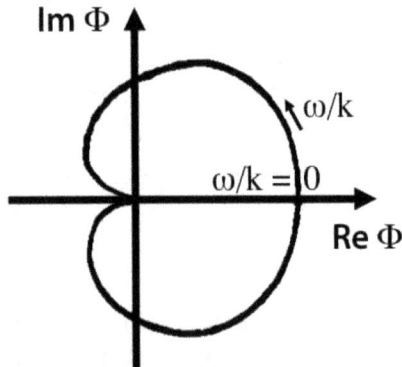

Fig. 4.1. The locus of $\Phi(\omega/k)$ in the complex plane as the phase speed, ω/k, is varied from $-\infty$ to ∞. The point corresponding to $\omega/k = 0$ is the intersection of the locus with the real Φ axis as indicated.

ity towards exponentially growing oscillations by the Nyquist criterion [i.e. that the hodograph for $\epsilon(\omega, k)$ not enclose the origin].

Equation (4.1) becomes

$$\varphi = q \int \frac{d^3 k}{(2\pi)^3} \frac{\exp(i\mathbf{k}\cdot\mathbf{x})}{k^2 + \Phi(\mathbf{k}\cdot\mathbf{v}_0)} . \tag{4.5}$$

This is the starting point for all the following calculations. We shall consider only test particle velocities much less than the mean electron velocity so that we may approximate $\Phi_e(\omega/k)$ to the static value $k_{De}^2 \equiv \Phi_e(0)$, the square of the (generalised) inverse Debye length for the electrons. Thus

$$\Phi(\omega/k) \approx k_{De}^2 + \omega_{pi}^2 \int_{-\infty}^{\infty} dv \frac{g_i'(v)}{\omega/k - v + i0} . \tag{4.6}$$

Note that if the electrons are extremely hot with respect to the ions then $k_{De}^2 \to 0$ and the ions form an essentially one component plasma with the electrons forming a neutralising background. The physical explanation of this is that, although the electrons are much lighter than the ions, they are moving too fast to be appreciably deflected from their paths within the range of the test particle and hence do not take part in the screening. If, on the other hand, the test particle has a velocity comparable with that of an average electron then the ions may be regarded as a uniform background due to their inertia. This case is therefore formally identical to the case $k_{De}^2 = 0$.

We give finally the normalized Maxwellian and Lorentzian distribution functions for a species s in a non-relativistic plasma together with the corresponding polarization functions Φ_s.

Maxwellian case:

$$g_s(v) = \left(\frac{m_s}{2\pi T_s}\right)^{1/2} \exp\left(-\frac{m_s v^2}{2T_s}\right) = \frac{k_{Ds}}{(2\pi)^{1/2}\omega_{ps}} \exp\left(-\frac{1}{2}\left[\frac{k_{Ds}v}{\omega_{ps}}\right]^2\right), \tag{4.7}$$

where m_s is the particle mass and T_s the temperature in energy units, giving

$$\Phi_s\left(\frac{\omega_{ps}}{k_{Ds}}x\right) = k_{Ds}^2\left[1 - \sqrt{2}\,x\exp\left(-\frac{x^2}{2}\right)\Psi\left(\frac{x}{\sqrt{2}}\right)\right.$$
$$\left. + i\left(\frac{\pi}{2}\right)^{1/2}x\exp\left(-\frac{x^2}{2}\right)\right], \tag{4.8}$$

for any dimensionless x, where $k_{Ds}^2 \equiv e_s^2 n_s/\varepsilon_0 T_s$ and

$$\Psi(y) \equiv \int_0^y dt\,\exp\left(t^2\right) . \tag{4.9}$$

Lorentzian case:

$$g_s(v) = \frac{1}{\pi} \frac{u_{s0}}{v^2 + u_{s0}^2} \ . \tag{4.10}$$

and

$$\Phi_s\left(\frac{\omega_{ps}}{k_{Ds}} x\right) = \frac{k_{Ds}^2}{(1 - ix)^2} \ , \tag{4.11}$$

where $k_{Ds} \equiv \omega_{ps}/u_{s0}$.

4.3. Analytical approximations

Although it is clearly impossible analytically to evaluate the integral Eq. (4.5) for general $g(v)$, or even for as simple a distribution as the Lorentzian, one may derive approximations for the integral in various ranges of v_0 that enable one to gain some understanding of its properties.

First we sketch $\Phi(\omega/k)$ and its asymptotic expansion in Fig. 4.2.

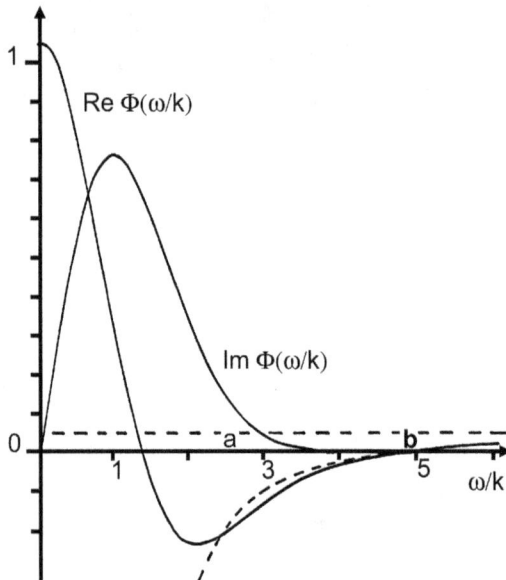

Fig. 4.2. The polarization function $\Phi(\omega/k)$ (in units such that $\omega_{pi} = k_{Di} = 1$) for a Maxwellian ion distribution and electron Debye constant $k_{De}^2 = 0.5$ (dashed horizontal line). The asymptotic expansion $k_{De}^2 - (k^2/\omega^2 + \overline{v_i^2}k^4/\omega^4)$ is included for comparison, being represented by the dashed curve.

4.3.1. *Small v_0*

We note from Eq. (4.5) that the argument of $\Phi \equiv \Phi_e + \Phi_i$ is less than or equal to v_0. Consequently for small v_0 only the behaviour of Φ near the origin will affect the potential.

Expanding about the origin, $\Phi(w) = k_D^2 + ia_0 w - a_1 w^2 + O(w^3)$, where a_0 and a_1 are positive constants depending on the distribution function, we get, in units such that $k_D^2 \equiv k_{Di}^2 + k_{De}^2 = 1$,

$$\frac{1}{k^2 + \Phi(w)} = \frac{1}{k^2 + 1} - \frac{ia_0 w}{(k^2 + 1)^2}$$
$$+ \left[\frac{a_1}{(k^2 + 1)^2} - \frac{a_0^2}{(k^2 + 1)^3} \right] w^2 + \dots . \qquad (4.12)$$

Then

$$\varphi(\mathbf{x}) = q \int \frac{d^3 k}{(2\pi)^3} e^{i\mathbf{k} \cdot \mathbf{x}} \left\{ \frac{1}{k^2 + 1} - ia_0 \frac{\hat{\mathbf{k}} \cdot \mathbf{v}_0}{(k^2 + 1)^2} \right.$$
$$\left. + \left[\frac{a_1}{(k^2 + 1)^2} - \frac{a_0^2}{(k^2 + 1)^3} \right] \mathbf{v}_0 \cdot \hat{\mathbf{k}} \hat{\mathbf{k}} \cdot \mathbf{v}_0 \right\}$$
$$= q \left\{ \frac{\exp(-r)}{4\pi r} - ia_0 \mathbf{v}_0 \cdot \hat{\mathbf{x}} \int \frac{d^3 k}{(2\pi)^3} \frac{\hat{\mathbf{x}} \cdot \hat{\mathbf{k}}}{(k^2 + 1)^2} e^{i\mathbf{k} \cdot \mathbf{x}} \right.$$
$$\left. + \mathbf{v}_0 \mathbf{v}_0 : \left[\int \frac{d^3 k}{(2\pi)^3} \hat{\mathbf{k}} \hat{\mathbf{k}} \, e^{i\mathbf{k} \cdot \mathbf{x}} \left(\frac{a_1}{(k^2 + 1)^2} - \frac{a_0^2}{(k^2 + 1)^3} \right) \right] \right\}, \qquad (4.13)$$

where $\hat{}$ denotes a unit vector.

The first integral in the second line of Eq. (4.13) may be evaluated as follows

$$\int \frac{d^3 k}{(2\pi)^3} \frac{\hat{\mathbf{x}} \cdot \hat{\mathbf{k}}}{(k^2 + 1)^2} e^{i\mathbf{k} \cdot \mathbf{x}} = \frac{1}{(2\pi)^2} \int_{-1}^{1} d\mu \int_{0}^{\infty} dk \frac{k^2}{(k^2 + 1)^2} e^{ik\mu r}$$
$$= \frac{1}{i(2\pi)^2} \Re \frac{d}{dr} \left(\frac{1}{r} - \frac{d}{dr} \right) \eta(r), \qquad (4.14)$$

where $\eta(r)$ is defined as follows

$$\eta(z) \equiv \frac{z}{i} \int_{0}^{\infty} dx \frac{\exp(ix)}{x^2 + z^2} \qquad (4.15)$$

for $|\arg z| < \pi/2$ and by analytic continuation elsewhere, cutting the complex plane along the negative imaginary axis. For details of the properties of η see Appendix I of the thesis [14].

The tensor term in the second line of Eq. (4.13) may be evaluated by contour integration, yielding

$$\varphi(\mathbf{x}) = q \left\{ \frac{\exp(-r)}{4\pi r} - a_0 \frac{\hat{\mathbf{x}} \cdot \mathbf{v}_0}{(2\pi)^2} \Re \frac{d}{dr} \left(\frac{1}{r} - \frac{d}{dr} \right) \eta(r) \right.$$

$$\left. + \mathbf{v}_0 \cdot \left[\mathcal{T}_{\parallel} \hat{\mathbf{x}} \hat{\mathbf{x}} + \mathcal{T}_{\perp} (\mathbf{I} - \hat{\mathbf{x}} \hat{\mathbf{x}}) \right] \cdot \mathbf{v}_0 \right\}, \qquad (4.16)$$

where

$$\mathcal{T}_{\parallel} \equiv \frac{1}{8\pi} \left\{ a_1 \left[\left(1 + \frac{2}{r} + \frac{4}{r^2} + \frac{4}{r^3} \right) e^{-r} - \frac{4}{r^3} \right] \right.$$

$$\left. - \frac{a_0^2}{4} \left[\left(r + 3 + \frac{8}{r} + \frac{16}{r^2} + \frac{16}{r^3} \right) e^{-r} - \frac{16}{r^3} \right] \right\}, \quad (4.17)$$

$$\mathcal{T}_{\perp} \equiv \frac{1}{8\pi} \left\{ a_1 \left[\frac{2}{r^3} - \frac{1}{r} \left(1 + \frac{2}{r} + \frac{2}{r^2} \right) e^{-r} \right] \right.$$

$$\left. - \frac{a_0^2}{4} \left[\frac{8}{r^3} - \left(1 + \frac{4}{r} + \frac{8}{r^2} + \frac{8}{r^3} \right) e^{-r} \right] \right\}. \qquad (4.18)$$

Note that, contrary to appearances, $\mathcal{T}_{\parallel}$ and $\mathcal{T}_{\perp}$ are regular at $r = 0$.

For large r, the asymptotic form of $\eta(r) \sim 1/r$ holds and the exponential terms may be neglected. Thus, for $r \to \infty$,

$$\varphi(\mathbf{x}) \sim q \left\{ \frac{a_0}{\pi^2} \frac{\hat{\mathbf{x}} \cdot \mathbf{v}_0}{r^3} + \frac{(a_0^2 - a_1)}{2\pi} \frac{(\hat{\mathbf{x}} \cdot \mathbf{v}_0)^2}{r^3} - \frac{(a_0^2 - a_1)v_0^2}{4\pi} \frac{[1 - (\hat{\mathbf{x}} \cdot \hat{\mathbf{v}}_0)^2]}{r^3} \right\}.$$
$$(4.19)$$

The first term is dominant for $|\hat{\mathbf{x}} \cdot \mathbf{v}_0| \gg 0$ i.e. for all directions not oblique to the direction of motion. The field is reminiscent of a dipole field except that it decays more rapidly with r. This may qualitatively be interpreted as meaning that the centre of the screening cloud has been displaced to a position behind the particle. It is this asymmetry that gives rise to the drag on a very heavy particle moving at subthermal speeds and, since this drag is the summation of the field of each individual particle moving in the average field of all the others, one supposes that the drag thus calculated includes single particle effects to the extent of validity of the linearised theory,[b] i.e. except for very close collisions.

Rand [17] has calculated the screened field to first order in v_0 by considering the trajectories of individual particles, explicitly rather than through the Vlasov equation. He obtains a function $\phi_1(x)$ as a triple integral, which

---

[b]This is confirmed by the work of Hubbard [11] who shows that the friction coefficient is the sum of the 'self field' term and a fluctuating microfield term which vanishes in the limit of infinite test-particle mass, as do the higher transition moments.

he apparently evaluates numerically but which agrees exactly[c] with values calculated from $\frac{1}{2}\Re(d/dx)(d/dx - 1/x)\eta(x)$.

It is interesting to note that the solution we have obtained differs completely from that of Pines and Bohm [16] and workers who have used linearised fluid dynamical equations [19, 21]. We ascribe this discrepancy to these authors' use of a collective approach in a region in which it is not valid: Inspection of Fig. 4.2 shows that the region in which collective coordinates exist can only be $|\omega/k| \gtrsim \bar{v}_i^2$, which is never entered in the case of a slow particle. The criterion $k \lesssim k_{Di}$ is not adequate in the case of forced oscillations since ω may be much less than ω_{pi}.

We note that behind the test particle the potential has opposite sign to its unscreened value, which we refer to as positive. At right angles to the direction of motion the first order term vanishes and the second-order term dominates. We see that $\varphi(\mathbf{x})$ goes negative in this direction if $a_0^2 > a_1$. For the Lorentzian distribution we have from Eq. (4.11) $\Phi_i(x) = 1 + 2ix - 3x^2 + k_{De}^2 + O(x^3)$, in units where $\omega_{pi} = k_{Di} = 1$. Thus $a_0 = 2$ and $a_1 = 3$, and therefore $a_0^2 > a_1$ so that the 90° potential does go negative somewhere between $r = 0$ and $r = \infty$.

We shall now discuss the situation when the particle has sufficient velocity for wave excitation to be possible.

The dispersion relation for plasma waves is $\epsilon(\omega, k) = 0$, i.e.

$$\Phi(\omega/k) = -k^2 , \tag{4.20}$$

where ω is in general complex, but if the waves are but slightly damped then we may take ω real as a first approximation. It is then clear that only the regions in which $\Re\Phi(\omega/k)$ is negative are available for excitation. In Fig. 4.2 this means the region $a < \omega/k < b$ which, it will be noted, exists if k_{De}^2 is sufficiently small.

This is the region of ion acoustic waves, which resemble sound waves in ordinary gases in that there is an upper bound, b, to their phase velocity. To guide our mathematical investigations we shall endeavour to give a physical picture of of the processes that can occur.

If a particle has a velocity v_0, $a < v_0 < b$, then it can excite waves with the same phase velocity as its own velocity. We might therefore expect it to be followed by a train of waves with this phase velocity, spreading out laterally with the maximum group velocity possible. This is roughly indicated in Fig. 4.3. If $v > b$ then we can expect no monochromatic train

---

[c]There is an error in Eq. (26) of Ref. 2, namely that the second Z in the equation should be replaced by $2Z/(Z + 1)$.

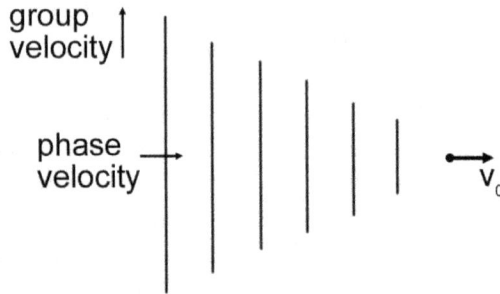

Fig. 4.3.   Schematic of wake field behind a charged test particle moving through a plasma.

but some kind of shock wave confined within the 'Mach cone'. This is analogous to the formation of the shock wave behind a supersonic object in the atmosphere or to a longitudinal Cerenkov radiation.

A note of caution must be sounded at this point with respect to the spreading out of the wake. Since this is controlled by the interference of the generated waves it is critically dependent on the dispersive nature of the medium, as will be discussed later. It is well known that in regions of anomalous dispersion in optical media the concept of group velocity breaks down entirely[28]. We thus expect the situation to be complicated if Landau damping becomes important.

Since we are seeking wave behaviour a contour integral method is the obvious choice and has been much used, see e.g. Ref. 16. We indicate the general method below as well as some of its analytical difficulties, not mentioned in the literature due in effect to the assumption from the outset that an asymptotic form for the dielectric constant is a valid approximation.

In Eq. (4.5) using cylindrical coordinates with axis parallel to $\mathbf{v}_0$ leads to

$$\varphi = \frac{q}{(2\pi)^2} \int_0^\infty dk_\perp \, k_\perp J_0(k_\perp x_\perp)$$

$$\times \int_{-\infty}^\infty dk_1 \exp(ik_1 x_1) \left[k_1^2 + k_\perp^2 + \Phi\left(\frac{k_1 v_0}{(k_1^2 + k_\perp^2)^{\frac{1}{2}}} \right) \right]^{-1}. \quad (4.21)$$

The contour in the k_1 integral may be completed in the lower/upper half of the complex k_1 plane according as $x_1 \lessgtr 0$. However, note that, since $k_1 = \pm k_\perp$ are branch points, the contours must be indented as shown

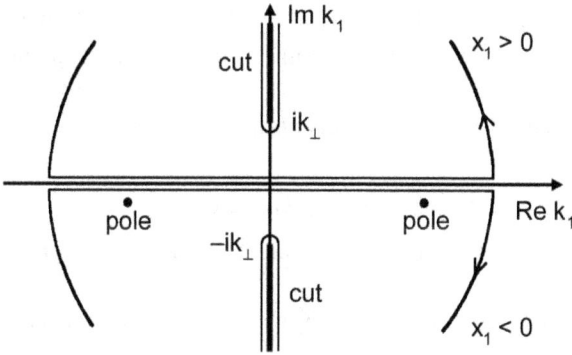

Fig. 4.4.   Contours used in the evaluation of the k_1 integral in Eq. (4.21).

in Fig. 4.4. The zeros of the denominator will give rise to poles whose contributions may be evaluated by the method of residues. For reasonably stable distributions, at least, one may show that the plasma wave poles corresponding to Eq. (4.20) are in the lower half k_1 plane. Consequently there is a wave excitation only behind the particle. There are other zeros of ϵ somewhere in the lower half plane, but since they give rise to strongly damped contributions, one hopes that they may be ignored, at any rate for large $|x_1|$. Indeed the contribution from the region of the cut will also be negligible compared with the wave contribution.

For $x_1 > 0$ it is necessary to consider the contribution from the region of the cut. The part nearest the real axis is clearly the most important for large $|x_1|$, so let us consider the region around $ik_\perp$

In this region $|k_1 v_0/(k_1^2 + k_\perp^2)^{1/2}|$ is large and the asymptotic expansion for Φ may be used. This may be obtained from Eq. (4.6) by expanding the denominator and integrating term by term, giving

$$\Phi(z) = k_{\text{De}}^2 - \omega_{\text{pi}}^2 \left(\frac{1}{z^2} + \frac{\overline{v_i^2}}{z^4} \right) - 2\pi i\, \theta(-\Im z)\, g_i'(z)\, \omega_{\text{pi}}^2\,, \qquad (4.22)$$

where the last term is required for analytical continuation into the lower half z plane. Here $\overline{v_i^2}$ is the mean-square ion speed (the square of the ion thermal speed in a Maxwellian plasma). However, if k_1 is in the lower/upper half plane, so also is $k_1/\left(k_1^2 + k_\perp^2\right)^{1/2}$. Hence the analytical continuation term is not required in the region we are considering. At $k_1 = -ik_\perp$ it gives rise in general to an essential singularity, but this contribution is being neglected as mentioned above. We now note the curious fact that the asymptotic expansion is single-valued in the upper half plane since only even powers of $\left(k_1^2 + k_\perp^2\right)^{1/2}$ occur. Thus there is only a simple pole in the neighbourhood of $k_1 = ik_\perp$.

For simplicity let us retain only terms to order $1/z^2$. Note that to this order the maximum phase velocity is

$$b = \omega_{\text{pi}}/k_{\text{De}} \equiv C_{\text{s}} , \qquad (4.23)$$

the constant C_{s} being commonly called the 'ion sound' speed.

4.3.2. Intermediate velocity $(\overline{v_i^2})^{1/2} \ll v_0 < C_{\text{s}}$ (forward field)

If $v_0 \ll b$ then the pole is approximately at $k_1 = ik_\perp$ and we may take $k_{\text{De}} = 0$ without greatly altering the situation. Then

$$\varphi = \frac{q}{4\pi} \left[\frac{1}{r} - \frac{\omega_{\text{pi}}}{v_0} \int_0^\infty dy \, \frac{J_0\left(\frac{\omega_{\text{pi}}x_\perp y}{v_0}\right) \exp\left(-\frac{\omega_{\text{pi}}x_1 y}{v_0}\right)}{1 + y^2} \right] . \qquad (4.24)$$

An alternative form is obtained by noting that [29]

$$\frac{1}{r} = \int_0^\infty dk_\perp \, J_0(k_\perp x_\perp) \exp(-x_1 k_\perp)$$

$$= \frac{\omega_{\text{pi}}}{v_0} \int_0^\infty dy \, J_0\left(\frac{\omega_{\text{pi}}x_\perp y}{v_0}\right) \exp\left(-\frac{\omega_{\text{pi}}x_1 y}{v_0}\right) . \qquad (4.25)$$

Thus

$$\varphi = \frac{q}{4\pi} \frac{\omega_{\text{pi}}}{v_0} \int_0^\infty dy \, \frac{y^2}{1 + y^2} J_0\left(\frac{\omega_{\text{pi}}x_\perp y}{v_0}\right) \exp\left(-\frac{\omega_{\text{pi}}x_1 y}{v_0}\right) . \qquad (4.26)$$

Note that the characteristic length is now v_0/ω_{pi}, the distance the particle moves in an ion plasma oscillation time, rather than the ion Debye length. The new length is longer, and furthermore, the screening is again not exponential at large r as we shall now show.

$$\text{For } x_\perp = 0 : \varphi = \frac{q}{4\pi} \frac{1}{x_1} \left[1 - \frac{\omega_{\text{pi}}x_1}{v_0} in\left(\frac{i\omega_{\text{pi}}x_1}{v_0}\right) \right]$$

$$\sim \frac{q}{4\pi} \frac{2\omega_{\text{pi}}}{v_0} \left(\frac{\omega_{\text{pi}}r}{v_0}\right)^{-3} . \qquad (4.27)$$

$$\text{For } x_1 = 0 : \varphi = \frac{q}{4\pi} \frac{1}{x_\perp} \left\{ 1 - \frac{\omega_{\text{pi}}x_\perp}{v_0} \left[I_0\left(\frac{\omega_{\text{pi}}x_\perp}{v_0}\right) - \mathbf{L}_0\left(\frac{\omega_{\text{pi}}x_\perp}{v_0}\right) \right] \right\}$$

$$\sim -\frac{q}{4\pi} \frac{\omega_{\text{pi}}}{v_0} \left(\frac{\omega_{\text{pi}}r}{v_0}\right)^{-3} , \qquad (4.28)$$

where I_0 and L_0 modified Bessel and Struve functions respectively.

A general asymptotic form at large r may be obtained from Eq. (4.26) by approximating the denominator to 1 and differentiating Eq. (4.25) twice with respect to $\omega_{\rm pi}x_1/v_0$. Thus

$$\varphi \sim \frac{q}{4\pi}\frac{\omega_{\rm pi}}{v_0}\frac{2(\hat{\mathbf{x}}\cdot\hat{\mathbf{v}}_0)^2-(\hat{\mathbf{x}}\times\hat{\mathbf{v}}_0)^2}{(\omega_{\rm pi}r/v_0)^3}. \tag{4.29}$$

Note that this obeys the same inverse third power law as in the very low velocity case, and also as in this case the field goes negative, for large r, at large angles between $\hat{\mathbf{x}}$ and $\hat{\mathbf{v}}_0$. The angle at which the field changes sign is approximately $55°$.

4.3.3. *Supersonic velocities $v_0 > C_{\rm s}$ (forward field)*

As v_0 increases beyond maximum phase velocity b (the ion sound speed) the pole moves from $ik_\perp$ to $ik_{\rm De}$ and the forward field changes to a Debye potential with Debye length $k_{\rm De}^{-1}$. This has the physical interpretation that if the particle velocity greatly exceeds the maximum ion wave phase velocity then the ions are too sluggish to participate in the screening.

We must now discuss the case $x_1 < 0$ where plasma wave excitation is assumed to be the dominant contribution. The positions of the plasma wave poles are given by the dispersion relation Eq. (4.20). It will be assumed that the asymptotic expansion Eq. (4.22) represents a valid approximation to Φ, with the one alteration that, since z is on or near the real axis, the last term is $-i\pi g_i'(z)$. [It will be noted that in a sector containing the real axis $g_i'(z)$ usually decays exponentially, and the factor multiplying it is irrelevant to the asymptotic expansion; but we require an approximation to Φ and so analytically continue off the real axis where the $i\pi$ factor is known to be exact.]

The dispersion relation is now

$$k^2 + k_{\rm De}^2 - \omega_{\rm pi}^2\left(\frac{k^2}{\omega^2}+\overline{v_{\rm i}^2}\frac{k^4}{\omega^4}\right) - i\pi\omega_{\rm pi}^2 g_i'\left(\frac{\omega}{k}\right) = 0. \tag{4.30}$$

This is valid for $|\omega/k| \gg (\overline{v_{\rm i}^2})^{1/2}$, i.e. when the 'finite temperature' term $v_{\rm i}^2 k^4/\omega^4$ is small compared with k^2/ω^2.

4.3.4. Intermediate velocity $(\overline{v_i^2})^{1/2} < v_0 < C_s$ (wake)

We again make the simplifying assumption that $k_{De} = 0$. The dispersion relation can now be reduced to the familiar form

$$\omega^2 \approx \omega_{pi}^2 + \overline{v_i^2}k^2 + i\pi\omega_{pi}^2 \operatorname{sgn}\omega \left(\frac{\omega_{pi}}{k}\right)^2 g_i'\left(\frac{\omega_{pi}}{k}\right) = 0 \,. \qquad (4.31)$$

To simplify the analysis let us use units such that $k_{Di} = \omega_{pi} = 1$. In these units $\overline{v_i^2} = 3$ (Maxwellian case), and generally we may assume $\overline{v_i^2} \sim 1$.

The finite temperature correction is small provided $k^2 \ll 1$, i.e. for wavelengths much greater than a Debye length. Thus, for the dispersion relation to be valid, we must require both $|k_1| \ll 1$ and $k_\perp \ll 1$. Since $\omega = k_1 v_0 \approx 1$ it is clear that we require $v_0 \gg 1$.

Substituting $\omega = k_1 v_0$ and $k^2 = k_1^2 + k_\perp^2$, and assuming $k_\perp$ small, we find

$$k_1 = \pm\frac{1}{v_0}\left\{1 + \frac{\alpha}{2} + \alpha v_0^2 k_\perp^2 \right.$$
$$\left. \pm\frac{i\pi}{2}\left[\frac{1+\alpha}{v_0^2} + k_\perp^2\right]^{-1} g'\left(\left[\frac{1+\alpha}{v_0^2} + k_\perp^2\right]^{-1/2}\right)\right\}, \quad (4.32)$$

where $\alpha \equiv \overline{v_i^2}/v_0^2 \ll 1$. The last term takes account of Landau damping and, $g'(v)$ being negative, serves to displace the poles in Fig. 4.4 below the real k_1 axis. Neglecting the damping term, which is very small until $k_\perp$ gets close to 1, gives

$$\frac{\partial}{\partial k_1}\left[k_1^2 + k_\perp^2 + \Phi\left(\frac{k_1 v}{(k_\perp^2 + k_1^2)^{1/2}}\right)\right]_{\text{poles}}$$
$$= \pm\frac{2}{v}\left(1 + \frac{\alpha}{2}\right)\left(1 + \frac{\alpha}{2}y^2\right)(1 + y^2)\,, \quad (4.33)$$

where $y \equiv k_\perp v_0$.

The two factors containing α are small correction terms. Using Eq. (4.32) to locate the poles and Eq. (4.33) to evaluate the residues we find from Eq. (4.21) that

$$\varphi = \frac{2-\alpha}{v_0} \Im\left[\exp\left(-i\left[1 + \frac{\alpha}{2}\right]\left|\frac{x_1}{v_0}\right|\right) I\left(\frac{\alpha}{2}i\left|\frac{x_1}{v_0}\right|, \frac{x_\perp}{v_0}, \frac{\alpha}{2}\right)\right], \quad (4.34)$$

where

$$\gamma(y) \equiv -\frac{\pi}{2}\frac{v_0^2}{(1+\alpha+y^2)} g'\left(\frac{v_0}{(1+\alpha+y^2)^{1/2}}\right) \qquad (4.35)$$

is the Landau damping term, which has the effect of rapidly cutting off the integral when $|y|$ gets close to v_0, and I is the integral

$$I(a, b, c) \equiv \int_0^\infty dy\, y\, \frac{J_0(by) \exp(-iay^2)}{(1 + y^2)(1 + cy^2)} \psi(y) , \qquad (4.36)$$

with

$$\psi(y) \equiv \exp\left(-\left|\frac{x_1}{v_0}\right| \gamma(y)\right) . \qquad (4.37)$$

In Eq. (4.34) the substitutions for the dummy arguments a, b and c are

$$a = \frac{\alpha}{2} \left|\frac{x_1}{v_0}\right| , \quad b = \frac{x_\perp}{v_0} \quad \text{and} \quad c = \frac{\alpha}{2} . \qquad (4.38)$$

When $b = 0$ we suppose that the cutoff factor ψ may be neglected. Then

$$I(a, 0, c) \approx \frac{1}{2(1 - c)} \left[e^{ia} E_1(ia) - e^{ia/c} E_1\left(\frac{ia}{c}\right)\right] \qquad (4.39)$$

where E_1 is the exponential integral [30].

To obtain an asymptotic form in the general case it is helpful to transform the one-sided integral into a two-sided integral by a method due to Hankel [29]. noting that, for $x > 0$,

$$J_0(x) = \frac{1}{2}\left[H_0^{(1)}(x) - H_0^{(1)}(-x + i0)\right] , \qquad (4.40)$$

where $H_0^{(1)}(z)$ is a Hankel function, defined on the complex z-plane cut along the negative real axis, being regular elsewhere. Then

$$I(a, b, c) = \frac{1}{2} \int_{-\infty}^\infty dy\, y\, \frac{H_0^{(1)}(b[y + i0]) \exp(-iay^2)}{(1 + y^2)(1 + cy^2)} \psi(y) . \qquad (4.41)$$

If a were zero and ψ well behaved, we could evaluate the integral by completing the contour in the upper half plane. Instead we seek the saddle points and poles of the integrand, the poles being at $y = \pm i$ and $y = \pm i/\sqrt{c}$. To find the saddle points let us suppose that the asymptotic form for $H_0^{(1)}(z)$,

$$H_0^{(1)}(z) \sim \sqrt{\frac{2}{\pi}} \frac{\exp i(z - \frac{\pi}{4})}{\sqrt{z}} , \quad |z| \to 0 , \qquad (4.42)$$

is valid in the region of these points. (The asymptotic form will be valid in the region of the poles provided $|b| \gg 1$, $c < 1$.)

Since the denominator is slowly varying away from its zeros the saddle points are determined by the exponential factor, the exponent of which, being quadratic, has only one stationary point. This is at $y = y_0$, where $y_0 = b/2a$. For Eq. (4.41) to be valid we require $|y_0| \gg 1$.

We conclude that, provided the two conditions $b \gg 1$ and $b \gg 2|a|$ are satisfied, we may approximate I by

$$I(a,b,c) \approx \frac{e^{-i\pi/4}}{\sqrt{b}} \sqrt{2\pi} \int_{-\infty}^{\infty} \frac{dy}{2\pi} \frac{(y+i0)^{1/2} \exp(-iay^2)\psi(y)}{(1+y^2)(1+cy^2)} e^{iby} . \quad (4.43)$$

This is a Fourier transform, which may be transformed by the convolution theorem to

$$I(a,b,c) \approx \frac{e^{-i\pi/4}}{\sqrt{b}} \frac{1}{\sqrt{2\pi}} \psi_b * \int_{-\infty}^{\infty} dy \frac{(y+i0)^{1/2} \exp(-iay^2)e^{iby}}{(1+y^2)(1+cy^2)} , \quad (4.44)$$

where $\psi_b \equiv \int_{-\infty}^{\infty} \psi(y)e^{iby}dy/2\pi$ is a delta-like function of b with width $1/v_0$, which will have the effect of destroying all wave structure with wavelength of a Debye length or less.

The integral may now be estimated by deforming the contour of integration to cross the saddle point along the line of steepest descent. The pole at $i/\sqrt{c}$, being far up the imaginary axis, gives negligible contribution. See Fig 4.5 for the contour used to evaluate the integral.

We make the further requirement that the width of the saddle at y_0 be much less than y_0. That is, $1/\sqrt{a} \ll y_0$. Then

$$I(a,b,c) \approx \psi_b * \left[\sqrt{\frac{\pi}{2b}} \frac{e^{ia-b}}{1-c} - \frac{i}{\sqrt{2ab}} \frac{y_0^{1/2} \exp(ib^2/4a)}{(1+y_0^2)(1+cy_0^2)} \right] . \quad (4.45)$$

Since the first term varies slowly compared with ψ_b, the convolution will leave it but little changed, whereas the rapid fluctuations in $\exp(ib^2/4a)$

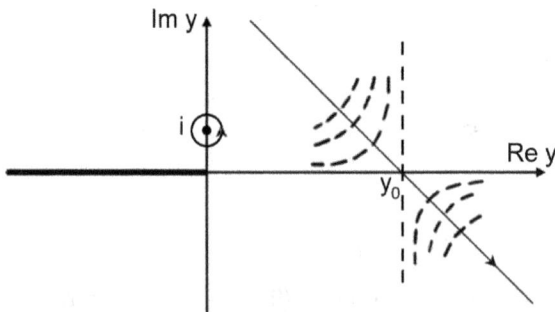

Fig. 4.5.   The contour used for estimating the integral in Eq. (4.44).

will be damped out. Supposing that the exponential term may locally be approximated by a monochromatic wave, with $y_0 = b/2a$, we finally obtain

$$I(a,b,c) \approx \sqrt{\frac{\pi}{2b}} \frac{e^{ia-b}}{1-c} - \frac{i}{\sqrt{2ab}} \frac{y_0^{1/2} \exp(ib^2/4a)}{(1+y_0^2)(1+cy_0^2)} \, . \qquad (4.46)$$

The method of steepest descent we have used to approximate the integral is equivalent, in the neighbourhood of the saddle point, to the method of stationary phase employed by Majumdar [20] but takes better account of the behaviour away from the saddle point by including the contribution of the pole at $y = i$. We have also attempted to take some account of Landau damping, which Majumdar was unable to do owing to his formulation in terms of fluid dynamics.

Below we recapitulate the criteria for the validity of Eq. (4.46),

$$c \ll 1, \quad b \gg 1, \quad b \gg 2a, \quad b \gg 2\sqrt{a} \, . \qquad (4.47)$$

With a, b and c given by Eq. (4.38) these correspond physically (recalling that in this subsection we are using units such that $k_{Di} = \omega_{pi} = 1$) to the inequalities

$$\frac{\alpha}{2} \ll 1, \quad \frac{x_\perp}{v_0} \gg 1, \quad x_\perp \gg \alpha|x_1|, \quad x_\perp \gg (\overline{v_i^2}|x_1|)^{1/2} \, . \qquad (4.48)$$

Observe that the Landau damping term cuts off the second term at about $x_\perp = |x_1|/v_0$, which might be called a thermal Mach cone. This behaviour can be understood from the fact that plasma waves with high group velocity

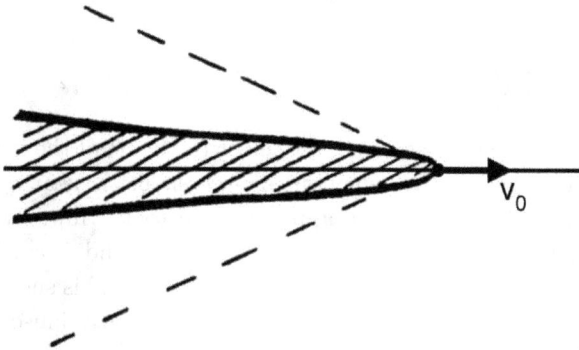

Fig. 4.6. Schematic of the wake region of an intermediate-velocity test particle indicating where the approximations used to derive Eq. (4.46) break down (hatched region) and the 'thermal Mach cone' (dashed line).

have low phase velocity, and hence are strongly damped. The first term, however, is unaffected by Landau damping and exhibits no thermal Mach cone, although it decays exponentially with $x_\perp$. We indicate the thermal Mach cone and the region excluded by the inequalities in Eq. (4.48) in Fig. 4.6.

4.3.5. Supersonic velocities $v_0 > C_s$ (wake field)

This is the case considered by Kraus and Watson [22] but their treatment neglects finite ion temperature effects on the dispersion relation and is consequently inadequate for finding variations over distances less than k_{De}^{-1}. We shall not go into detail on this case but shall indicate the approximations that may be made and the integral obtained.

The simplifying assumption made by the above authors can be represented as

$$(\overline{v_i^2})^{1/2} \ll b \ll v_0 , \tag{4.49}$$

where b is here the upper bound to the phase velocity [$\approx C_s$ by Eq. (4.23)], not the dummy argument used in the previous subsection. Taking units such that $\omega_{pi} = k_{De} = 1$ we have

$$\overline{v_i^2} \ll 1 , \quad b \approx 1 , \quad v_0 \gg 1 \quad \text{and} \quad k_{Di} \gg 1 . \tag{4.50}$$

Using the same methods as before, but neglecting Landau damping, we find the approximate form for $x_1 < 0$

$$\varphi \approx -\frac{2}{v_0} \int_0^\infty \frac{dk_\perp \, k_\perp^2}{(1 + k_\perp^2)^{3/2}} \, J_0(x_\perp k_\perp) \sin\left(\left|\frac{x_1}{v_0}\right| \frac{k_\perp(1 + \frac{1}{2}\overline{v_i^2}k_\perp^2)}{(1 + k_\perp^2)^{1/2}}\right) . \tag{4.51}$$

Kraus and Watson in effect assume that the critical part of the integral for determining the large $|x_1|/v_0$ behaviour of ($x_\perp$ small) is near $k_\perp = 0$ so that terms such as $1 + k_\perp^2$ may be approximated to 1. However, it must be observed that there is a range of $k_\perp$ (between 1 and $1/\overline{v_i^2}$) in which the argument of the sine function is but slowly varying. This suggests that, superimposed on the endpoint contribution, there may be a sinusoidal term.

When $x_\perp$ is not small we may transform the integral into a two-sided integral as in Sec. 4.3.4. There will be a saddle point which disappears into the origin at the Mach cone $x_\perp = |x_1|/v_0$, but the contribution from the singularities of the integrand will be complicated by the essential singularities at $k_\perp = \pm i$.

4.4. Numerical solution

The explorations of the previous section showed that some insight into the nature of the fields around a particle in a collisionless plasma, as given formally by Eq. (4.1), may be obtained by asymptotic analyses in various limits. However, these are often difficult and the errors difficult to quantify. It is thus essential to supplement such analytic work with numerical and graphical methods (and *vice versa*).

In this section we summarise how this was done computationally in my MSc thesis work [14] using the newly arrived IBM 7044 (which had just replaced the Australian-built computer CSIRAC [31]).

4.4.1. *Formulation of the numerical method*

To evaluate the integral Eq. (4.1) numerically we must choose axes such that as much of the integration as possible may be done analytically while the remaining integrations are as tractable as possible. It is found that, although one may come tantalisingly close to reducing the triple integral to a single integral, it is in general necessary to evaluate a double integral.

Three systems of coordinates suggest themselves: the cylindrical coordinates of Sec. 4.3, spherical polars with axis along $\mathbf{v}_0$, and spherical polars with axis along $\mathbf{x}$. The first method involves infinite integrals with inte-

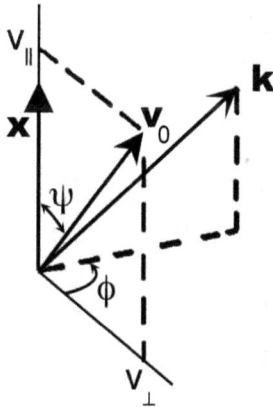

Fig. 4.7. Coordinate system used for numerical evaluation of the triple integral in Eq. (4.1).

grands having a large peak near the plasma wave pole and we consequently reject it. The last choice appears to have the advantage over the remaining one that it is easier to understand the behaviour of the integrands (at least at low $\mathbf{v}_0$) and to isolate the singularity caused by the behaviour of $\eta(x)$ at $x = 0$. These axes are sketched in Fig. 4.7. In this coordinate system we have

$$\hat{\mathbf{k}}{\cdot}\mathbf{v}_0 = \mu v_\| + \sqrt{1 - \mu^2}\,\cos\phi\,v_\perp \, , \tag{4.52}$$

where $v_\| \equiv \mathbf{v}_0{\cdot}\hat{\mathbf{x}} = v_0\sin\psi$, $v_\perp \equiv |\mathbf{v}_0{\times}\hat{\mathbf{x}}| = v_0\cos\psi$, and $\mu \equiv \hat{\mathbf{k}}{\cdot}\hat{\mathbf{x}}$.

In these coordinates Eq. (4.1) becomes

$$\varphi = \frac{q}{(2\pi)^3} \int_{-1}^{1} d\mu \int_0^{2\pi} d\phi \,\Re \int_0^{\infty} dk\, \frac{k^2 \exp(ik\mu r)}{k^2 + \Phi(\mu v_\| + \sqrt{1 - \mu^2}\,\cos\phi\,v_\perp)} \, , \tag{4.53}$$

where $r \equiv |\mathbf{x}|$ and Φ is defined in Eq. (4.3). The units used were such that $k_{\mathrm{Di}} = \omega_{\mathrm{pi}} = 1$ and $q = 4\pi$.

By using the fact that $\Phi(x)^* = \Phi(-x)$ for real x, the range of the μ and ϕ integrations was reduced by half. It was found useful [14] to define the new special function $\eta(z)$, Eq. (4.15). Analytical properties of $\eta(z)$ are discussed in Appendix A of the thesis [14] and a listing of a Fortran IV subroutine for its efficient evaluation is given. This was used in the code developed to calculate the results presented below. (More discussion of the numerical method is given in the thesis [14].)

In terms of η, Eq. (4.53) reduces to a double integration

$$\varphi = \frac{2}{\pi^2 r} \int_0^{\pi} d\phi \left\{ \frac{\pi}{2} + \int_0^{r} dx\, \Im\left[\sqrt{\Phi}\,\eta(x\sqrt{\Phi}) \right] \right\} \, , \tag{4.54}$$

where

$$\sqrt{\Phi} \equiv \left\{ \Phi\left(\frac{x v_\|}{r} + \left[1 - \left(\frac{x}{r}\right)^2 \right]^{1/2} \cos\phi\, v_\perp \right) \right\}^{1/2} \, . \tag{4.55}$$

4.4.2. *Small v_0 results*

The case where the test particle is travelling slower than the ion thermal speed was studied using the Lorentzian distribution function Eq. (4.10) with $k_{\mathrm{De}} = 0$. The treatment of Sec. 4.3.1 indicates that the choice of distribution function and electron Debye length simply alters the relative magnitude of the constants k_{D}, a_0 and a_1, so we expect there is little to be gained by study of a large variety of cases.

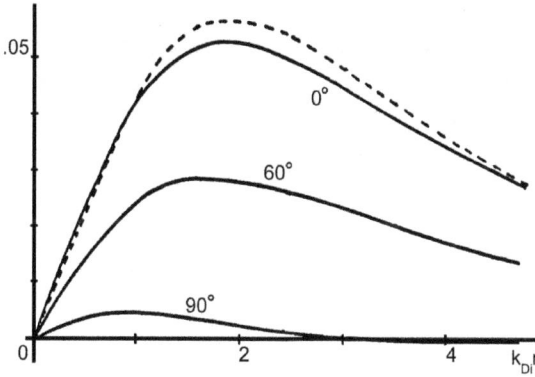

Fig. 4.8. Solid curves are the computed deviation from the Debye potential, as defined in the text, for a particle moving at velocity $v_0 = 0.2\omega_{\rm pi}/k_{\rm Di}$ through a plasma with Lorentzian ion distribution and infinite-temperature electrons. The dashed curve is an analytic approximation to the forward field from Eq. (4.16).

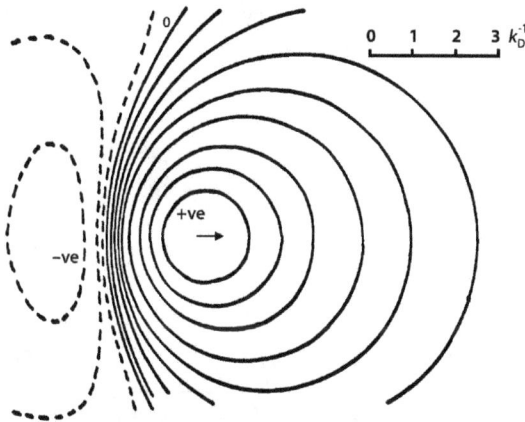

Fig. 4.9. Equipotentials in a plasma the same as assumed for Fig. 4.8 but with $v_0 = 0.4\omega_{\rm pi}/k_{\rm Di}$ (the direction of $\mathbf{v}_0$ is indicated by an arrow). The solid equipotentials indicate positive (+ve) φ and the dashed lines negative (−ve). The scale in Debye lengths is indicated top right.

In Fig. 4.8 we plot the deviation from the Debye potential along lines making angles ψ of $0°$, $60°$ and $90°$ to the direction of travel, the deviation being defined as $[\varphi(\mathbf{x}) - \varphi_{\rm D}(r)]/\varphi_0(r)$, where $\varphi_{\rm D}(r) \equiv q\exp(-k_{\rm Di}r)/4\pi r$ is the Debye potential and $\varphi_0(r) \equiv q/4\pi r$ is the bare, unscreened potential.

The approximate result at $0°$ obtained from the term of Eq. (4.16) first order in v_0 is also plotted (dashed curve), the agreement being seen to be quite reasonable.

The qualitative form of Fig. 4.9 is also in agreement with that predicted from the analytical work. Note the potential well behind the particle, beyond which the field becomes attractive to charges of the same sign as the test particle.

4.4.3. *Intermediate v_0 results*

The intermediate case, where the test particle moves faster than the ion thermal speed but less than the ion sound speed C_s, was studied in part using both Lorentzian and Maxwellian distribution functions as these to some extent represent extremes of Landau damping. The Lorentzian distribution is sufficiently simple for an exact dispersion relation to be derivable, thus obviating the use of asymptotic expansions of Φ. Again we took $k_{De} = 0$, so $C_s = \infty$ by Eq. (4.23).

Potentials on the axis of symmetry, parallel (forward field, $\psi = 0$) or antiparallel (wake field, $\psi = \pi$) to the direction of motion of the test particle, are plotted in Figs. 4.10, 4.11, and 4.12.

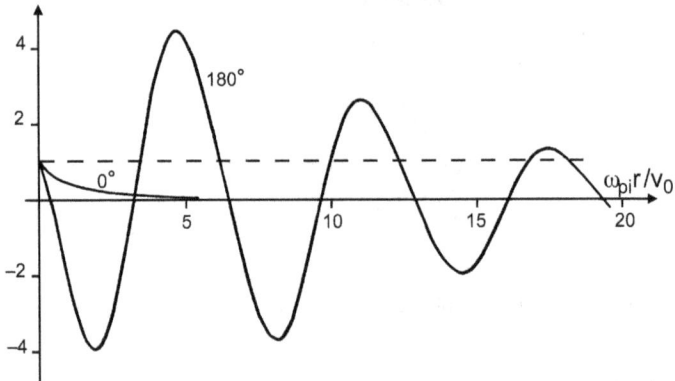

Fig. 4.10.   Plots of $\varphi(r, \psi)/\varphi_0(r)$ as a function of distance r (in units of v_0/ω_{pi}) calculated for a test particle moving at $v_0 = 8\omega_{pi}/k_{Di}$ in a plasma with Lorentzian ions and $k_{De} = 0$. The forward field is labelled $0°$ and the wake field is labelled $180°$. The dashed line is the unscreened case, $\varphi = \varphi_0$.

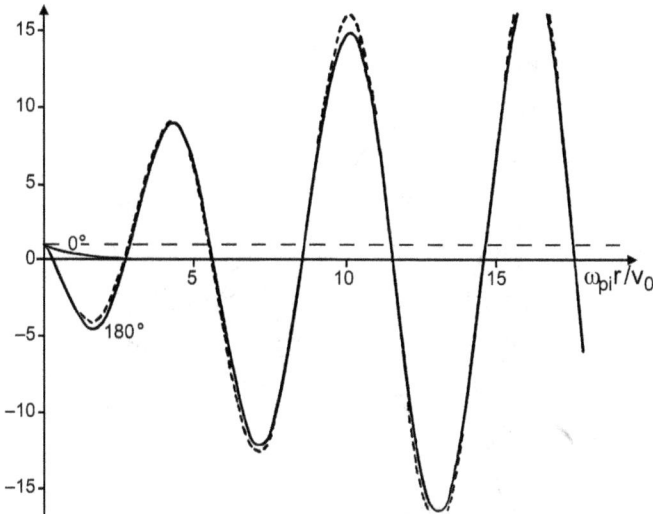

Fig. 4.11. Plots of $\varphi(r, \psi)/\varphi_0(r)$ as a function of distance r (in units of v_0/ω_{pi}) calculated for a test particle moving at $v_0 = 8\omega_{\mathrm{pi}}/k_{\mathrm{Di}}$ in a plasma with Maxwellian ions and $k_{\mathrm{De}} = 0$. The forward field is labelled $0°$ and the wake field is labelled $180°$. The short-dashed curve is the approximation Eq. (4.39).

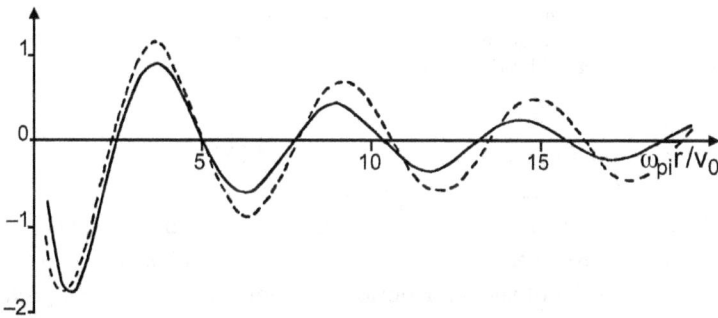

Fig. 4.12. Same case as in Fig. 4.11 except that the test particle speed is halved, $v_0 = 4\omega_{\mathrm{pi}}/k_{\mathrm{Di}}$. Plotted is $[\varphi(r, \psi)/\varphi_0(r)](\omega_{\mathrm{pi}}r/v_0)^{-1}$ in the wake, $\psi = \pi$.

- *Forward field:*

  In all three of the plots the results agreed well with Eq. (4.27), with agreement particularly good for the Maxwellian case (four significant figures at $v_0 = 8\omega_{\mathrm{pi}}/k_{\mathrm{Di}}$. [The potentials at $90°$ to the direction of

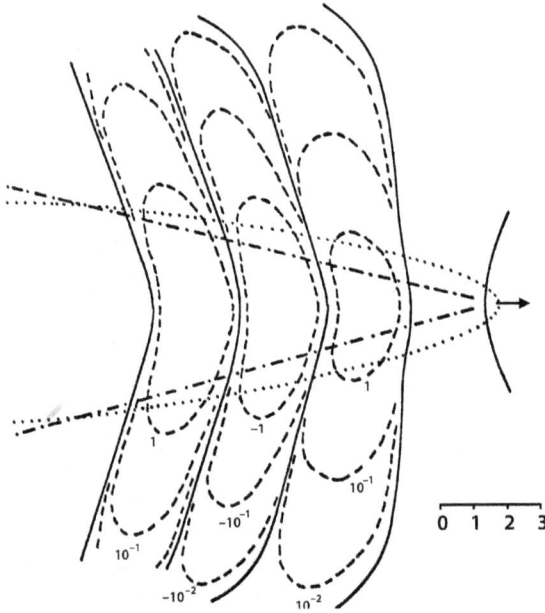

Fig. 4.13. Contours of $\varphi(\mathbf{x})/\varphi_0(r)$ for a test particle moving with $v_0 = 4\omega_{\mathrm{pi}}/k_{\mathrm{Di}}$ (the direction of $\mathbf{v}_0$ is indicated by an arrow) in a plasma with Maxwellian ions and $k_{\mathrm{De}} = 0$. The scale in units of v_0/ω_{pi} is indicated lower right, the 'thermal Mach cone' $x_\perp = |x_1|\omega_{\mathrm{pi}}/k_{\mathrm{Di}}v_0$ is indicated by dash-dotted lines, and the boundary of the region of validity of Eq. (4.46) is indicated by the dotted curve (cf. Fig. 4.6).

motion also agreed reasonably with Eq. (4.28).]

- *Wake field, Lorentzian case*, see Fig. 4.10:

  Within the wake Landau damping had a dominant effect in the Lorentzian case even at the centre of the wake. This is due to the slow decay of the tail of the distribution function. The analytical prediction (which was not derived in Sec. 4.3) was virtually indistinguishable on the scale of the graph, which is to be expected because the dispersion relation is exact.

- *Wake field, Maxwellian case*:

  For a Maxwellian distribution function the agreement with Eq. (4.39) was quite good in the $v_0 = 8$ case, Fig. 4.11, but not so good in the $v_0 = 4$ case, Fig. 4.12. This may be explained by the inadequacy of the asymptotic expansion Eq. (4.22) when $|\omega/k| < 4$. Landau damping may also have some effect.

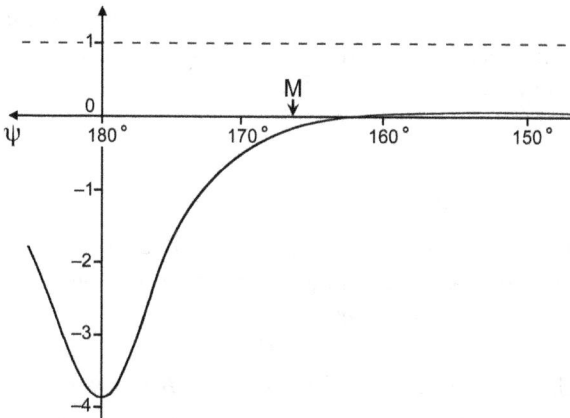

Fig. 4.14. Plot of $\varphi(r,\psi)/\varphi_0(r)$ vs. ψ (with $r = 48/k_{\mathrm{Di}}$) in the wake of a supersonic test particle moving at $v_0 = 32\omega_{\mathrm{pi}}/k_{\mathrm{Di}} = 4C_{\mathrm{s}}$ in a plasma with $k_{\mathrm{De}} = k_{\mathrm{Di}}/8$. Surprisingly, no shock wave is seen at the expected position of the Mach cone, which is calculated from $x_\perp/|x_1| = \omega_{\mathrm{pi}}/v_0 k_{\mathrm{De}}$ and indicated by the arrow marked M at $\psi = 166°$.

Owing to computing time limitations (at the time these cqalculations were made), the complete wake structure was mapped only in the $v_0 = 4$ Maxwellian case. A contour plot is given in Fig. 4.13, which represented almost two hours of computing time on the IBM 7044 and is still not very accurate.

4.4.4. *Supersonic case*

The supersonic case, $v_0 > C_{\mathrm{s}}$, considered in Sec. 4.3.5 was only briefly studied numerically. Fig 4.14 shows a fairly sharp cutoff near the Mach cone but no shock front. Indeed the function decays monotonically off the axis in complete contrast with the Kraus–Watson [22] result, which predicts an initial increase.

4.4.5. *Conclusion*

In the intervening years since my MSc work other authors have made similar calculations. I have not attempted a complete literature search, but note that the research project must have been topical at the time as the inverse third power asymptotic behaviour in Eq. (4.19) was announced a year later by Montgomery, Joyce, and Sugihara [32], in a paper that has been cited 58 times. In fact the problem is even more topical today with the rise of interest in dusty plasmas, as shown by the paper by Ishihara and Vladimirov [33] on the wake potential of a dust grain in a plasma with ion flow being cited more than 80 times.

Acknowledgments

I am grateful to Ken Hines for providing an ambiance in which we research students were able to develop intellectually both through his gentle guidance and through mutual interactions. I am indebted particularly to Norm Frankel for indoctrinating me in statistical physics and kinetic theory and sharing his thoughts on many topics, and to Andrew Prentice for many stimulating conversations and for providing the subroutine I adapted for calculating the response function Φ.

References

[1] K. C. Hines, *Phys. Rev.* **97**, 1725 (1955).
[2] L. D. Landau, *J. Phys. (U.S.S.R.)*, **8**, 201, (1944).
[3] U. Fano, *Phys. Rev.* **92**, 328 (1953).
[4] S. Chandrasekhar, *Rev. Mod. Phys.* **15**, 1 (1943).
[5] R. Balescu, *Statistical Mechanics of Charged Particles*, Monographs in Statistical Physics, Wiley Interscience, (1963).
[6] D. C. Montgomery and D. A. Tidman, *Plasma Kinetic Theory*, Advanced Physics Monograph Series, McGraw-Hill, New York, (1964).
[7] N. E. Frankel, K. C. Hines, and R. L. Dewar, *Phys. Rev. A*, **20**, 2120 (1979).
[8] R. Balescu, *Physics of Fluids*, **3**, 52 (1960).
[9] A. Lenard, *Annals of Physics*, 390 (1960).
[10] W. B. Thompson and J. Hubbard, *Rev. Mod. Phys.* **32**, 714 (1960).
[11] J. Hubbard, *Proc. Roy. Soc. (London)* **A260**, 114 (1961).
[12] J. Hubbard, *Proc. Roy. Soc. (London)* **A261**, 371 (1961).
[13] N. Rostoker, *Phys. Fluids* **7**, 479 (1964).
[14] R. L. Dewar, *Particle-Field Interactions in a Plasma*, M.Sc. thesis, University of Melbourne, (unpublished) (1967).
[15] R. L. Dewar, *Aust. J. Phys.* **30**, 533 (1977).
[16] D. Pines and D. Bohm, *Phys. Rev.* **13**, 338 (1952).
[17] S. Rand, *Phys. Fluids* **2**, 649 (1959).
[18] S. Rand, *Phys. Fluids* **3**, 265 (1910).
[19] S. K. Majumdar, *Proc. Phys. Soc.* **76**, 657 (1960).
[20] S. K. Majumdar, *Proc. Phys. Soc.* **82**, 669 (1963).
[21] M. H. Cohen, *Phys. Rev.* **123**, 711 (1963).
[22] L. Kraus and K. M. Watson, *Phys. Fluids* **1**, 480 (1958).
[23] H. A. Pappert, *Phys. Fluids* **3** 966 (1960).

[24] A. V. Gurevich, *Geomagn. i Aeronomiya (USSR)* **4**, 3 (1964). [English transl. *Geomagn. and Aeronomy (USA)* **4**, 1 (1964)]

[25] U. M. Panchenko and L. P. Pitayevsky, *Geomagn. i Aeronomiya (USSR)*, **4** 256 (1964). [English transl. *Geomagn. and Aeronomy (USA)* **4**, 637 (1964)]

[26] A. V. Gurevich and L. P. Pitayevsky, *Geomagn. i Aeronomiya (USSR)* **4**, 817 (1964). [English transl. *Geomagn. and Aeronomy (USA)* **4**, 637 (1964)]

[27] A. V. Gurevich and L. P. Pitayevsky, *Phys. Rev. Letters* **15** 346 (1965).

[28] J. A. Stratton, in *Electromagnetic Theory* McGraw-Hill, New York, (1941) p. 333.

[29] G. N. Watson, in *A Treatise on the Theory of Bessel Functions*, Cambridge University Press, Cambdridge, U.K. (1944).

[30] M. Abramowitz and I. A. Stegun, *Handbook of Mathematical Functions.* Applied Mathematics Series **55**, National Bureau of Standards, U.S. Government Printing Office, Washington D.C. (1972).

[31] A. Moffat, in *Fifty Years of Computing at The University of Melbourne*, Department of Computer Science and Software Engineering Department of Information Systems Web Site, *http://www.cs.mu.oz.au/ alistair/fifty-years/mof06history.pdf*, (2006).

[32] D. Montgomery, G. Joyce, and R. Sugihara, *Plasma Physics* **10**, 681 (1968).

[33] O. Ishihara and S. V. Vladimirov, *Phys. Plasmas* **4**, 69 (1997).

Chapter 5

Self-Avoiding Walks as a Canonical Model of Phase Transitions

A. J. Guttmann

ARC Centre of Excellence for Mathematics and Statistics
of Complex Systems,
Department of Mathematics and Statistics,
The University of Melbourne, Victoria 3010, Australia
E-mail: tonyg@ms.unimelb.edu.au

The self-avoiding walk model has long been the most commonly cited model describing the behaviour of long-chain polymers in dilute solution. More recently, with the inclusion of monomer-monomer, monomer-surface or other forms of interactions, some physically and biologically important phenomena can be modelled. We describe a number of such situations.

Additionally, we consider in some detail a canonical model of phase transitions that arises when we consider a restricted class of self-avoiding walks (SAW) which start at the origin $(0, 0)$, end at (L, L), and are entirely contained in the square $[0, L] \times [0, L]$ on the square lattice $\mathbb{Z}^2$. The number of distinct walks is known to grow as $\lambda^{L^2+o(L^2)}$. I estimate $\lambda = 1.744550 \pm 0.000005$ as well as obtaining strict upper and lower bounds, $1.628 < \lambda < 1.782$. Associating a fugacity, x, with each step of the walk gives rise to a canonical model of a phase transition. For $x < 1/\mu$ the average length of a SAW grows as L, while for $x > 1/\mu$ it grows as L^2. Here μ is the growth constant of unconstrained SAW in two dimensions. At $x = 1/\mu$ we find that the average walk length grows as $L^{4/3}$.

5.1. Introduction

I consider several models of a phase transition that arise from the study of self-avoiding walks (SAW) with various interactions and subject to various geometrical constraints. These are shown to give rise to phase transitions, usually second order. The most well known and widely studied such model is the collapse transition, in which the SAW goes from an expanded phase to a collapsed phase. This is described in the following subsection, and in the subsequent subsection we describe the situation of a polymer in the presence of an adsorbing surface, and refer briefly to the more complex case in which both the above forms of interaction are present.

I then review a number of applications to biological problems, and finally devote most of our attention to recent work on the problem of SAW which start at the origin $(0, 0)$, end at (L, L), and are entirely contained in the square $[0, L] \times [0, L]$ on the square lattice $\mathbb{Z}^2$. Associating a fugacity, x, with each step of the walk gives rise to a canonical model of a phase transition.

5.1.1. *The collapse transition*

One of the best-known transitions that self-avoiding walks (SAW) undergo is the transition from the expanded phase, which occurs in a good solvent— a situation typical of a non-interacting SAW— to the collapsed phase, characteristic of a polymer in a poor solvent. This is sometimes referred to as a "coil-globule" transition in which the SAW goes from the expanded, coil phase with fractal dimension $d_f = 1/\nu < d$, where d is the dimension in which the SAW is embedded, and ν is the (non-interacting) exponent characterising the end-to-end length, $\langle R^2 \rangle_n \sim \text{const.} \times n^{2\nu}$ (see Fig. 5.1) to a collapsed globule like phase with fractal dimension d.

The situation may be modelled by introducing a fugacity x between nearest-neighbour monomers that are not adjacent monomers of the walk. The situation is shown in Fig. 5.2. Let $c_{m,n}$ denote the number of n-step

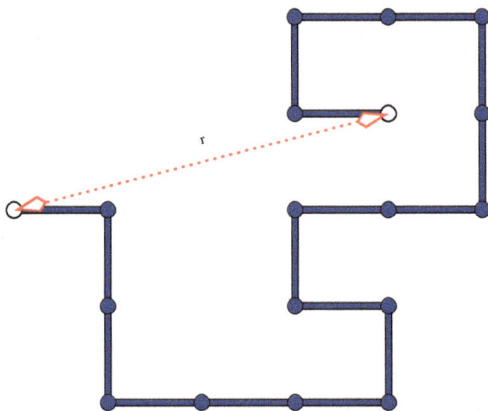

Fig. 5.1. A two-dimensional SAW, showing the Euclidean end-to-end distance.

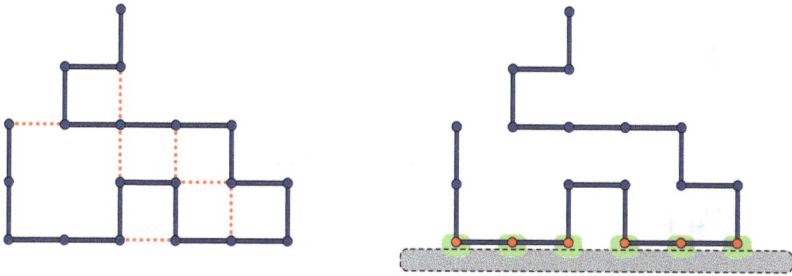

Fig. 5.2. An interacting SAW (left panel), showing monomer-monomer interactions (dotted). A SAW interacting with a surface (right panel) showing monomer-surface interactions (shaded).

SAW with m nearest neighbour contacts. Then we can form the generating function

$$C(x,y) = \sum_m \sum_n c_{m,n} x^m y^n.$$

By varying x we can tune the SAW to be attractive or repulsive. The fugacity dependent mean square end-to-end length exponent $\nu(x)$, defined by

$$\langle R^2(x)\rangle_n = \frac{\sum r^2 c_{m,n} x^m}{\sum c_{m,n} x^m} \sim \text{const.} \times n^{2\nu(x)} ,$$

where r is the Euclidean end-to-end distance of each SAW, changes discontinuously with x as shown in Fig. 5.3. In the repulsive, non-interacting or weakly attractive regime, $\nu(x)$ takes its non interacting value (respectively $3/4$, 0.58687 and $1/2(\log)$ in 2, 3 and 4 spatial dimensions). The result in two dimensions is believed to be exact, the result in three dimensions is a numerical estimate [1] believed accurate to a few places in the last quoted digit, and the result in four dimensions is again exact, up to a (known) confluent logarithmic correction. As x is tuned to become more attractive, the SAW collapses into a ball, characterised by exponent $\nu = 1/d$, where d is the spatial dimension. At a particular value of x, known as the θ-temperature, the exponent ν takes an intermediate value. This is known to be $\nu_\theta = 4/7$ in two dimensions. The situation is shown in Fig. 5.3.

In three dimensions $\nu_\theta = 1/2(\log)$, because for interacting SAW, three is the so-called 'upper critical dimension'. Again, there are confluent logarithmic terms. In four dimensions for many years there was some confusion as to the situation. Many earlier numerical studies found the transition to

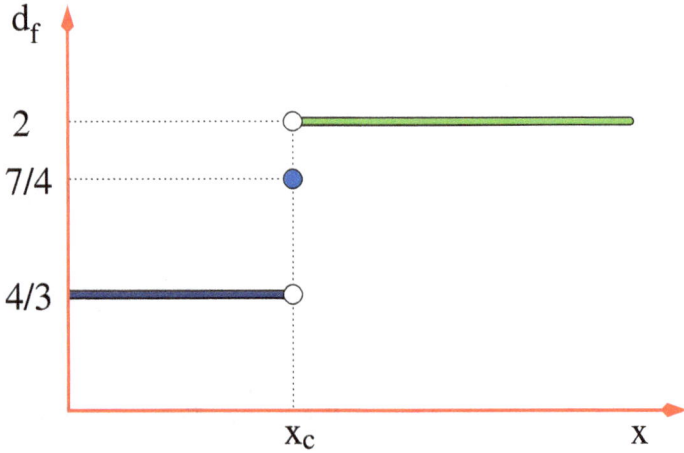

Fig. 5.3. The variation of fractal dimension d_f with interaction fugacity for a two-dimensional SAW.

behave like a rounded, first-order transition, rather than the second-order transition we have described in dimensions 2 and 3. Careful numerical and theoretical work by Prellberg and Owczarek [2] has resolved the situation. There is a distinct collapse transition at the θ-temperature, at which temperature the exponent takes the expected Gaussian value $1/2$ without any logarithmic correction. However for any finite polymer length, the transition has many characteristics of a rounded first-order transition. They explain how the pseudo-first-order transition is scaled away in the thermodynamic, or large length limit, leaving the mean-field or Gaussian second-order transition, as one would expect above the upper critical dimension. This happens as the latent heat decays (algebraically) to zero with polymer length.

5.1.2. The adsorption transition

In this situation there is a fugacity associated with monomers on a surface or, in two dimensions, a line. In the repulsive or neutral or weakly attractive regime, only a finite number (and hence a vanishingly small fraction) of monomers lie in the surface. This is the desorbed regime. In the regime where the fugacity is strongly attractive, a finite fraction of monomers lie in the surface. This is the adsorbed regime. The two-dimensional situation is shown in the right hand panel of Fig. 5.2. For three dimensional SAW with a two-dimensional adsorbing surface, the fractal dimension changes from its three-dimensional value $d_f \approx 1.704$ to the two-dimensional value $d_f = 4/3$. In two dimensions, with a one-dimensional adsorbing line, the fractal dimension changes from the two-dimensional value $d_f = 4/3$ to the one-dimensional value $d_f = 1$, while in four dimensions the desorbed-adsorbed transition is marked by a change in fractal dimension from $d_f = 2(\log)$ to $d_f = 1.704$. Unlike the collapse transition, there is no analogue of the θ point at which there is an intermediate fractal dimension. As the surface attraction increases, the fractal dimension remains that of the desorbed phase down to, and at, the transition temperature. Further surface attraction then results in a switch to the adsorbed phase.

The two types of interaction can be combined, so that one has monomer-monomer interactions as well as monomer-surface interactions. It has recently been shown [3] that, for a three-dimensional system with a two-dimensional surface, there is an infinite hierarchy of layering transitions in the attractive, or poor solvent regime. For finite length polymers there is a series of (rounded) layering transitions which increase in number and prominence with increasing polymer length. Krawczyk *et al.* [3] give compelling arguments and numerical evidence for the infinite length limit, resulting in a transition from a collapsed, but not a macroscopically adsorbed, state to a collapsed fully adsorbed state.

In the next section we briefly discuss the modelling of some biological phenomena by self-avoiding walks. These primarily involve the monomer-monomer or monomer surface interactions we have just described.

5.2. Biological models: SAW as models of DNA

By constraining self avoiding walks in various geometries, and introducing interactions by appropriate fugacities, SAW turn out to be useful in modelling a surprising variety of biological situations. In this section we will consider just three.

One of the first, historically, is the modelling of the denaturation of DNA. When a solution of DNA is heated, the double stranded molecules denature into single strands. In this process, "looping out" of AT rich regions of the DNA segments first occurs, followed eventually by separation of the two strands as the paired GC segments denature. This denaturation process corresponds to a phase transition [4].

A simple model of this DNA denaturation transition was introduced in 1966 by Poland and Scheraga [5, 6] (hereinafter referred to as PS) and refined by Fisher [7]. The model consists of an alternating sequence (chain) of straight paths and loops, which idealise denaturing DNA, as a sequence of double stranded and single stranded molecules. An attractive energy is associated with paths. Interactions between the different parts of a chain and, more generally, all details regarding real DNA such as chemical composition, stiffness or torsion, are ignored. It was found that the phase transition is determined by the critical exponent c of the underlying loop class. Due to the tractability of the problem of random loops, (rather than self-avoiding loops), that version of the problem was initially studied by PS [6]. The model displays a continuous phase transition in both two and three dimensions. It was argued by Fisher [7] that replacing random loops by self-avoiding loops, suggested as a more realistic representation accounting for excluded volume effects within each loop, sharpens the transition, but does not change its order. We shall investigate this particular model in a little more detail later.

Another early example of a biological problem modelled by SAW—or in this case self-avoiding polygons (SAP), was the study of vesicles by Leibler and Fisher [8], Fisher *et al.* [9], and Banavar *et al.* [10]. A vesicle is a biological object such as a blood cell whose behaviour is mediated by pressure. If the internal pressure exceeds the external pressure the cell will be inflated, whereas in the opposite situation it will be collapsed. Clearly, there will be a phase transition that occurs at a critical value of the pressure. Early series studies were largely restricted to two dimensions, the cell was modelled by a two-dimensional SAP, and the pressure induced phase transition was mediated by associating a fugacity with the enclosed area. Thus by varying this fugacity, polygon shapes of minimal area, which are long and thin, could be achieved, as could those of maximal area, which are effectively square. We will consider this problem in a little more detail later.

The third, and final, example is the use of SAW to model the micromanipulation of polymer molecules, particularly DNA, attached to a surface. In this situation, optical tweezers [11, 12] are used to pull the adsorbed biological molecule from the surface. This force is applied perpendicular to the adsorbing surface and will favour desorption. It is reasonable to expect some sort of a phase transition. At low levels of the force, the polymer remains adsorbed, but at higher levels it will be desorbed. There will be a temperature dependent force $f_c(T)$ between these two states. The shape of the force-temperature curve is of considerable interest, and can be con-

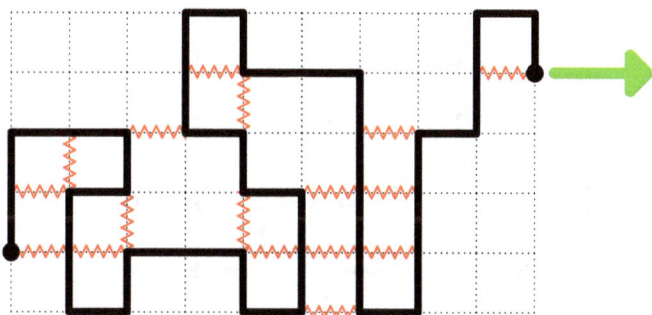

Fig. 5.4. A SAW model of a polymer subject to an elongation force.

sidered a phase boundary in the $T - f$ plane. This can be modelled by a SAW, tethered to a wall, with a fugacity associated with nearest-neighbour bonds, subject to a force perpendicular to the wall, as shown in Fig. 5.4.

5.2.1. *The DNA denaturation transition*

With the advent of efficient computers, it has recently become possible to simulate analytically intractable models extending the PS class, which are assumed to be more realistic representations of the biological problem. One of these is a model of two self-avoiding and mutually avoiding walks, with an attractive interaction between the different walks at corresponding positions in each walk [13–16]. This situation is illustrated in Fig. 5.5.

The model exhibits a first order phase transition in two and three dimensions. The critical properties of the model are described by an exponent c' related to the loop length distribution [14–19]. For PS models, this exponent coincides with the loop class exponent c if $1 < c < 2$. Within a refined model, where different binding energies for base pairs and stiffness are taken into account, the exponent c' seems to be largely independent of the specific DNA sequence and of the stiffness of paired walk segments corresponding to double stranded DNA parts [14]. There are, however, no simulations of melting curves for known DNA sequences which are compared to experimental curves for this model.

Some care needs to be taken in studying the literature on this problem, which may be misleading. This is discussed in considerable detail in Richard and Guttmann [20]. Loop classes discussed in the early approaches [6, 7] are classes of rooted loops and lead to chains which are not self-avoiding. This seems unsatisfactory from a biological point of view, since real DNA

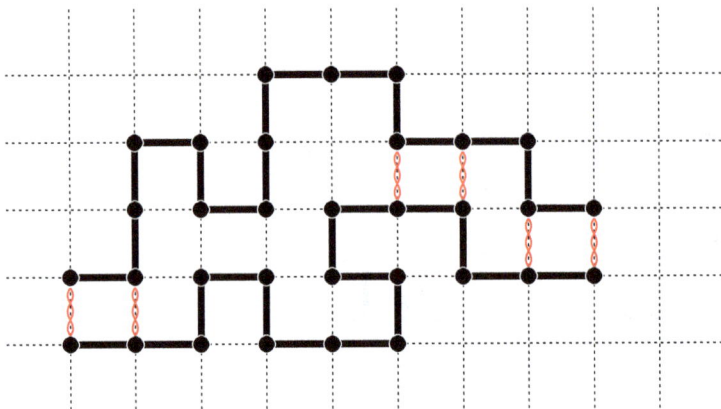

Fig. 5.5. A two-dimensional SAW model of DNA denaturation. Interactions between corresponding monomers on DNA strands are shown.

is self-avoiding. Secondly, the common view holds that PS models with self-avoiding loops cannot display a first order transition in two or three dimensions. In fact, this view led to extending the PS class [13, 17–19] in order to find a model with a first order transition. However this view is incorrect, as demonstrated [20] by a self-avoiding PS model with self-avoiding loops. Thirdly, the two exponents c and c', are used in the literature without distinction, although there are subtle differences.

Causo *et al.* [13] introduced a SAW model of this transition by considering pairs of SAW on the simple cubic lattice, with a common origin, which are allowed to overlap only at the same monomer position along each chain. That is to say, if one numbers the monomers from the common origin $0, 1, 2, \ldots i, \ldots n$, the two chains are mutually and individually self-avoiding except possibly where site i of the first monomer coincides with site i of the second monomer. A two dimensional version of this situation is illustrated in Fig. 5.6.

Such overlaps are encouraged by a fugacity ϵ, associated with such contacts. As the temperature increases, there is a transition temperature T_m above which the entropic advantage in breaking such bonds overcomes the attractive energy. From extensive simulations, it was concluded [13] that the transition is first order, as the energy density at the transition was found to be discontinuous.

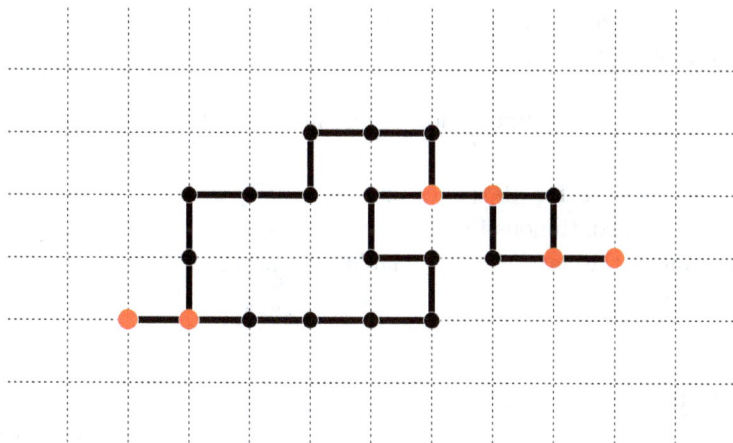

Fig. 5.6.   A two-dimensional SAW model of DNA denaturation. Interacting monomers are indicated with larger blobs.

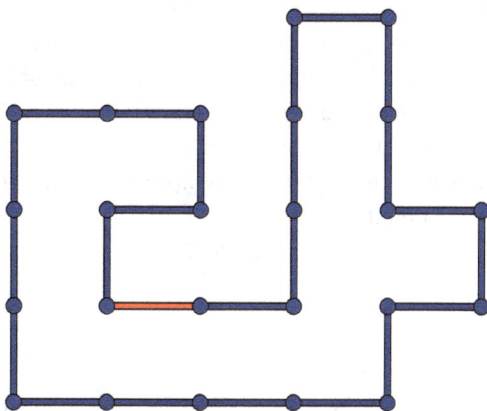

Fig. 5.7.   A two-dimensional SAP, showing its construction from a SAW.

5.2.2. *Vesicle collapse*

Vesicles are closed, solid objects, possessing a surface. A variant of SAW, in which the origin and end-point are adjacent, and are then joined up, is called a self-avoiding polygon (SAP). This situation is shown in Fig. 5.7.

In modelling vesicle collapse by SAP, let $p_{m,n}$ be the number of SAP per site on an infinite lattice, with perimeter m enclosing area n. Fisher *et al.* [9] proved that the free energy

$$\lim_{m \to \infty} \frac{1}{m} \log \sum_n p_{m,n} q^n := \kappa(q) \qquad (5.1)$$

exists and is finite for all values of the fugacity $q \leq 1$. Further, $\kappa(q)$ is log-convex and continuous for these values of q and is infinite for $q > 1$.

In terms of the natural two-variable generating function

$$P(x,q) = \sum_{x,q} p_{m,n} x^m q^n, \qquad (5.2)$$

it was further proved that for $q < 1$, $P(x,q)$ converges for $x < e^{-\kappa(q)}$, while for $q > 1$, $P(x,q)$ converges only for $x = 0$. The expected phase diagram is shown in Fig. 5.8. In the region below the phase boundary, the polygons are ramified objects, closely resembling branched polymers. That is to say, they are collapsed and string-like. As q approaches unity, they fill out more, and become less string-like. At $q = 1$ one has pure SAP. For $q > 1$ the polygons approximate squares, with their average area scaling as the square of their perimeter. Rigorous upper and lower bounds to the shape of the phase boundary have been found [9], and the locus of the actual phase boundary was found numerically from extrapolation of SAP enumerations by area and perimeter. In the extended phase $q = 1$, the mean area of polygons $\langle a \rangle_m$ of perimeter m grows asymptotically like $m^{3/2}$, whereas it grows like m in the deflated phase $q < 1$. It can be shown that in the

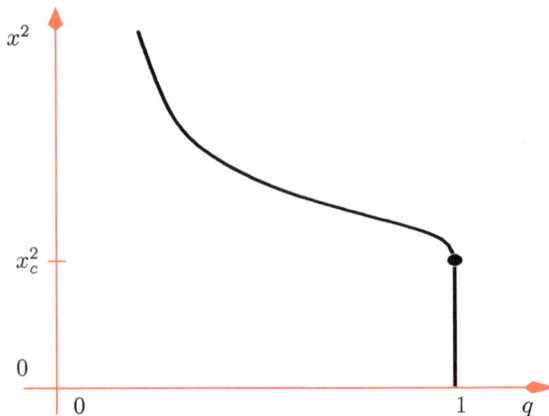

Fig. 5.8. The phase diagram showing the phase boundary $x_c(q)$.

limit $q \to 0$ the generating function is dominated by polygons of minimal area. Since for SAPs these polygons may be viewed as branched polymers, the phase $q < 1$ is also referred to as the branched polymer phase. This change of asymptotic behaviour is reflected in the singular behaviour of the perimeter and area generating function. Typically, the line $q = 1$ is a line of finite essential singularities for $x < x_c$. The line $x_c(q)$, where $P(x, q)$ is singular for $q < 1$, is typically a line of logarithmic singularities. For branched polymers in the continuum limit, the logarithmic singularity has been proved recently [21].

Of special interest is the point $(x_c, 1)$ where these two lines of singularities meet. The behaviour of the singular part of the perimeter and area generating function about $(x_c, 1)$ is expected to take the special form

$$P(x, q) \sim P^{(reg)}(x, q) + (1 - q)^\theta \, F[(x_c - x)(1 - q)^{-\phi}] \; ;$$
$$(x, q) \to (x_c^-, 1^-), \qquad (5.3)$$

where $F(s)$ is a scaling function of combined argument $s = (x_c - x)(1 - q)^{-\phi}$, commonly assumed to be regular at the origin, and θ and ϕ are critical exponents. The singular behaviour about $q = 1$ at the critical point x_c is then given by $P^{(sing)}(x_c, 1) \sim (1 - q)^\theta F(0)$. This scaling assumption implies an asymptotic expansion of the scaling function of the form

$$F(s) = \sum_{k=0}^\infty \frac{f_k}{s^{(k-\theta)/\phi}}. \qquad (5.4)$$

The leading asymptotic behaviour characterises the singularity of the perimeter generating function via $P(x, 1) \sim f_0(x_c - x)^{-\gamma}$, where $\theta + \phi\gamma = 0$. The first singularity of $F(s)$ on the negative axis determines the singularity along the curve $x_c(q)$. The locus on the axis (say at $s = s_c$) determines the line $x_c(q) \sim x_c - s_c(1 - q)^\phi$ near $q = 1$, which meets the line $q = 1$ vertically for $\phi < 1$.

Banavar *et al.* [10] studied a similar model analytically using a mapping onto a gauge model. As in other studies, they found the critical behaviour to be governed by a branched polymer fixed point. More recently, Richard, Guttmann and Jensen [22] have given a very persuasive conjecture as to the exact nature of the scaling function at the bi-critical point $(x_c^2, 1)$. In their work, it was natural to work with rooted SAP, and in that case the conjectured form of the scaling function was found to be

$$F^{(r)}(s) = \frac{x_c}{\pi\sigma} \frac{d}{ds} \log \mathrm{Ai} \left(\frac{\pi}{x_c} (2E_0)^{\frac{2}{3}} s \right) \qquad (5.5)$$

with exponents $\theta = 1/3$ and $\phi = 2/3$, where $Ai(x)$ is the Airy function. The conjectured form of the scaling function is then obtained by integration and is

$$F(s) - \frac{1}{12\pi}(1-q)\log(1-q) = -\frac{1}{\pi\sigma}\log \text{Ai}\left(\frac{\pi}{x_c}(2E_0)^{\frac{2}{3}}s\right) \qquad (5.6)$$

with exponents $\theta = 1$ and $\phi = 2/3$. The second term on the left-hand side is a constant of integration. The parameters for the square lattice are $\sigma = 2$ and $x_c = 0.379052277757(5)$. The parameters for the hexagonal lattice are $\sigma = 2$ and $x_c = 1/\sqrt{2 + \sqrt{2}}$ (known exactly [23]) and for the triangular lattice $\sigma = 1$ and $x_c = 0.2409175745(3)$. Further details of this conjecture have been published [24].

5.2.3. *Macromolecular desorption from a surface*

As briefly described above, in this situation a force is applied perpendicular to an adsorbing surface to which a polymer chain is attached. At low temperature, surface attraction dominates, but at high temperatures entropy dominates, and the polymer is free of the surface. The temperature dependent force needed to extend the polymer is calculated. Let the polymer have N monomers, of which n lie in the surface. (In two dimensions the "surface" is a line). Let $c_N(n, z)$ be the number of such SAW whose end-point is at perpendicular distance z from the surface. The model may be described by the partition function

$$Z_N(\omega, u) = \sum_{n,z} c_{n,z}\omega^n u^z \qquad (5.7)$$

where $\omega = e^{-\epsilon/kT}$, $u = e^{f/kT}$, $\epsilon < 0$ is the attractive energy of a monomer with the surface, and f is the force acting on the endpoint monomer, in a direction perpendicular to the surface.

Exactly solvable models based on Dyck paths and Motzkin paths in two dimensions, and a partially directed walk model in three dimensions have been studied [25], and re-entrant behaviour in three dimensions observed, but they were not in two. "Re-entrant behaviour" refers to the fact that the force temperature diagram at first increases with increasing temperature, and then decreases. It thus has a finite maximum at a positive temperature. Re-entrant behaviour is shown in Fig. 5.9.

The results of enumeration of interacting SAWs in two dimensions for $N \leq 30$ and in three dimensions for $N \leq 19$ have been found [26] giving a re-entrant force-temperature diagram in three dimensions, again but

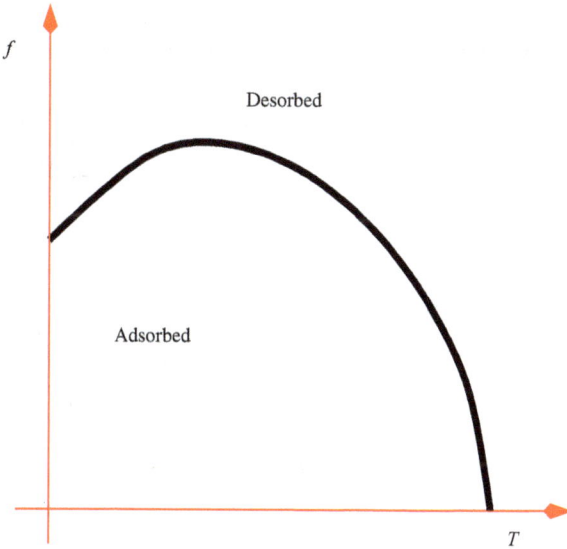

Fig. 5.9.   A re-entrant force-temperature curve.

not in two, exactly as observed [25] in the toy models. Based on much longer enumerations in two dimensions, to $N = 50$, Kumar *et al.* [27] obtained clear evidence of a re-entrant force-temperature diagram even in two dimensions. The "stick-release" behaviour of SAW pulled from a surface is also clearly shown in that study. In Figs. 5.10 and 5.11 the average length is plotted against applied force. The series of plateaus become smoothed out as the length of the walk increases at fixed temperature, and also as the temperature is raised for SAW of fixed length.

It can be seen from the above studies that SAW models are capable of shedding light on a wide variety of phase transitions associated with biological problems. The range of problems that can be addressed in this way seems to be limited only by the imagination of the researchers.

5.3. Self-avoiding walks crossing a square

We now consider the problem of self-avoiding walks on the square lattice $\mathbb{Z}^2$. For walks on an infinite lattice, it is generally accepted [28] that the number of such walks of length n, equivalent up to a translation, denoted c_n, grows as $c_n \sim \text{const.} \times \mu^n n^{\gamma-1}$, with metric properties, such as mean-square radius of gyration or mean-square end-to-end distance growing as

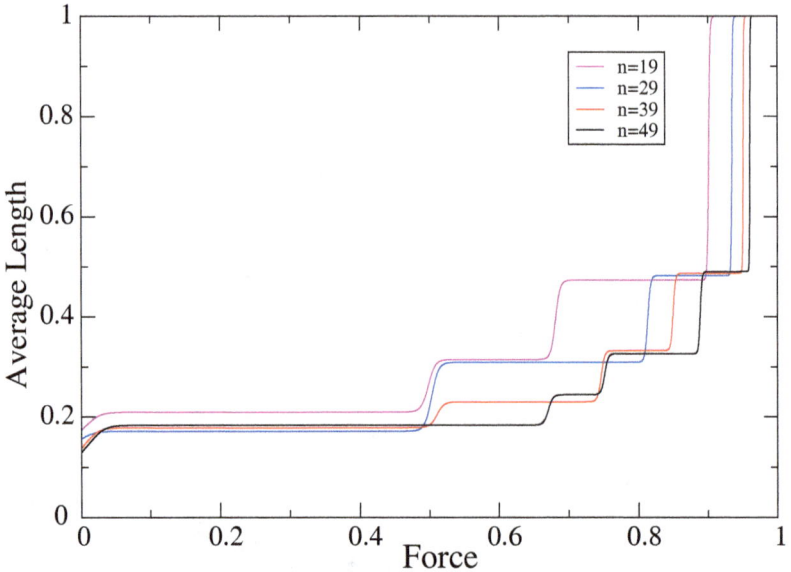

Fig. 5.10.   The "stick-release" behaviour of the force-extension curve for walks of various lengths at $T = 0.1$. Lengths increase from the left- to the right-hand curves.

$\langle R^2 \rangle_n \sim$ const. $\times n^{2\nu}$, where $\gamma = 43/32$ and $\nu = 3/4$. The growth constant μ is lattice dependent, and for the square lattice is not known exactly, but is indistinguishable numerically from the unique positive root of the equation $13x^4 - 7x^2 - 581 = 0$. I denote the generating function by $C(x) := \sum_n c_n x^n$. It will be useful to define a second generating function for those SAW which start at the origin $(0,0)$ and end at a given point (u, v), as $G_{(0,0;u,v)}(x)$. In terms of this generating function, the mass, $m(x)$, is defined [28] to be the rate of decay of G along a coordinate axis,

$$m(x) := \lim_{n \to \infty} \frac{-\log G_{(0,0;n,0)}(x)}{n}. \qquad (5.8)$$

Here, we are interested in a restricted class of square lattice SAW which start at the origin $(0,0)$, end at (L, L), and are entirely contained in the square $[0, L] \times [0, L]$. A fugacity, or weight, x is associated with each step of the walk. Historically, this problem seems to have led two largely independent lives. One as a problem in combinatorics (in which case the fugacity has been implicitly set to $x = 1$), and one in the statistical mechanics literature where the behaviour as a function of fugacity x has been

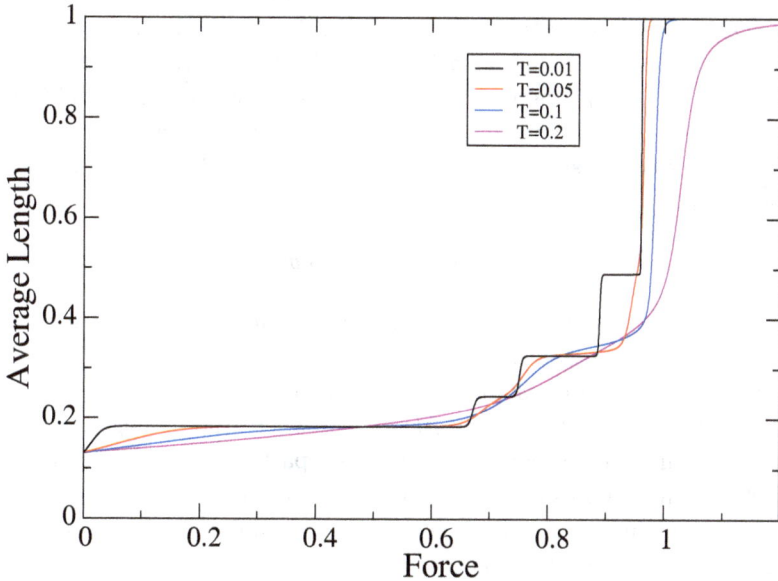

Fig. 5.11. The "stick-release" behaviour of the force-extension curve for 50 step walks at different temperatures. Temperatures increase from the left- to the right-hand curves.

of considerable interest, as there is a fugacity dependent phase transition. This problem has been thoroughly investigated very recently by Bousquet-Mélou, Guttmann and Jensen [29] and it is that work we describe below.

The problem seems to have first been seriously studied as a mathematical problem by Abbott and Hanson [30] in 1978, many of whose results and methods are still powerful today. A key question considered both then and now, is the number of distinct SAW on the constrained lattice, and their growth as a function of the size of the lattice. Let $c_n(L)$ denote the number of n-step SAW which start at the origin $(0,0)$, end at (L,L), and are entirely contained in the square $[0,L] \times [0,L]$. Further, let $C_L(x) := \sum_n c_n(L)x^n$. Then $C_L(1)$ is the number of distinct walks from the origin to the diagonally opposite corner of an $L \times L$ lattice. In Ref. [30] and independently [31], it was proved that $C_L(1) \sim \text{const.} \times \lambda^{L^2}$. The value of λ is not known, though bounds and estimates have been established [30, 31].

In the statistical mechanics literature, the problem appears to have been introduced by Whittington and Guttmann [31], who were particularly interested in the phase transition that takes place as one varies the fugacity associated with the walk length. All walks on lattices up to 6×6 were enumerated, and the estimate $\lambda = 1.756 \pm 0.01$ was given. At a critical value, x_c the average walk length of a path on an $L \times L$ lattice changes from $\Theta(L)$ to $\Theta(L^2)$, where $\Theta(x)$ is defined as follows: Let $a(x)$ and $b(x)$ be two functions of some variable x. We write that $a(x) = \Theta(b(x))$ as $x \to x_0$ if there exist two positive constants κ_1 and κ_2 such that, for x sufficiently close to x_0,

$$\kappa_1 \, b(x) \le a(x) \le \kappa_2 \, b(x).$$

The critical fugacity was estimated, and conjectured to be $x_c = 1/\mu$ [31]; a conjecture that was proved by Madras [32].

Abbott and Hanson [30] considered the slightly more general problem of SAW constrained to an $L \times M$ lattice, where the analogous question was asked: how many non-self-intersecting paths are there from $(0,0)$ to (L, M)? If one denotes the number of such paths by $C_{L,M}$, it is clear that, for M finite, the paths can be generated by a finite dimensional transfer matrix, and hence that the generating function is rational [33]. Indeed, it was proved [30] that

$$G_2(x) = \sum_{L \ge 0} C_{L,2} x^L = \frac{1 - x^2}{1 - 4x + 3x^2 - 2x^3 - x^4}. \tag{5.9}$$

It follows that $C_{L,2} \sim \text{const.} \times \lambda_2^{2L}$, where $\lambda_2 = \sqrt{\frac{2}{\sqrt{13}-3}} = 1.81735\ldots$

Following [31], Madras [32] proved a number of theorems. In fact, most of Madras's results were proved for the more general d-dimensional hypercubic lattice, but here we will quote them in the more restricted two-dimensional setting.

Theorem 5.1. *The following limits,*

$$\mu_1(x) := \lim_{L \to \infty} C_L(x)^{1/L} \quad \text{and} \quad \mu_2(x) := \lim_{L \to \infty} C_L(x)^{1/L^2},$$

are well-defined in $\mathbf{R} \cup \{+\infty\}$.

More precisely,

(i) *$\mu_1(x)$ is finite for $0 < x \le 1/\mu$, and is infinite for $x > 1/\mu$. Moreover, $0 < \mu_1(x) < 1$ for $0 < x < 1/\mu$ and $\mu_1(1/\mu) = 1$.*
(ii) *$\mu_2(x)$ is finite for all $x > 0$. Moreover, $\mu_2(x) = 1$ for $0 < x \le 1/\mu$ and $\mu_2(x) > 1$ for $x > 1/\mu$.*"

The average length of a (weighted) walk is defined to be

$$\langle n(x)\rangle_L := \sum_n nc_n(L)x^n / \sum_n c_n(L)x^n. \tag{5.10}$$

Theorem 5.2. *For* $0 < x < 1/\mu$, *we have that* $\langle n(x)\rangle_L = \Theta(L)$ *as* $L \to \infty$, *while for* $x > 1/\mu$, $\langle n(x)\rangle_L = \Theta(L^2)$.

The situation at $x = 1/\mu$ is unknown but there is compelling numerical evidence [29] that $\langle n(1/\mu)\rangle_L = \Theta(L^{1/\nu})$, where $\nu = 3/4$, in accordance with an intuitive suggestion [32].

Theorem 5.3. *For* $x > 0$, *define* $f_1(x) = \log \mu_1(x)$ *and* $f_2(x) = \log \mu_2(x)$.

(i) *The function* f_1 *is a strictly increasing, negative-valued convex function of* $\log x$ *for* $0 < x < 1/\mu$, *and* $f_1(x) = \Theta(-m(x))$ *as* $x \to 1/\mu^-$, *where* $m(x)$ *is the mass, defined by* (5.8).
(ii) *The function* f_2 *is a strictly increasing, convex function of* $\log x$ *for* $x > 1/\mu$, *and satisfies* $0 < f_2(x) \le \log \mu + \log x$.

Some, but not all of the above results were proved previously [31], but these three theorems elegantly capture all that is rigorously known.

5.3.1. *Bounds on the growth constant* λ

For the more general problem of SAW going from $(0,0)$ to (L, M) on an $L \times M$ lattice, it was proved [30] that, for each fixed M, $\lim_{L \to \infty} C_{L,M}^{\frac{1}{LM}} = \lambda_M$ exists. It immediately follows that $\lim_{L \to \infty} C_{L,L}^{\frac{1}{L^2}} = \lambda$ exists, which was also proved earlier in a somewhat different manner [31].

5.3.1.1. *Upper bounds on* λ

Following the work of Abbott and Hanson [30], consider any non-intersecting path crossing the $L \times L$ square. Label each unit square in the $L \times L$ lattice by 1 if it lies to the right of the path, and by 0 if it lies to the left. This provides a one-to-one correspondence between paths and a subset of $L \times L$ matrices with elements 0 or 1. Matrices corresponding to allowed paths are called admissible, otherwise they are inadmissible. Since the total number of $L \times L$ 0 − 1 matrices is 2^{L^2}, we immediately have the weak bound $C_{L,L} \le 2^{L^2}$. Of the 16 possible 2×2 matrices, only 14 can correspond to portions of non-intersecting lattice paths. Note that there are only 12 actual paths from

$(0, 0)$ to $(2, 2)$, but a further two matrices may correspond to paths that are embedded in a larger lattice. This leads to the bound $C_{L,L} \leq 14^{(L/2)^2}$, so $\lambda \leq 1.9343...$ Similarly, for 3×3 lattices there are 320 admissible matrices (out of a possible 512), so $\lambda \leq 320^{1/9} = 1.8982..$ For 4×4 lattices, Abbott and Hanson [30] claim that there are 22662 admissible matrices, but we believe the correct number to be 22816, giving the bound $\lambda \leq 1.8723...$ My colleagues and I have made dramatic extensions of this work, using a combination of finite-lattice methods and transfer matrices [29], and have determined the number of admissible matrices up to 19×19. There are $3.5465202... \times 10^{90}$ such matrices, giving the bound

$$\lambda \leq 1.781684.$$

A geometric interpretation of these matrices has been given [29] in terms of families of mutually avoiding SAW going from one boundary of the square to another. The same bound results of course.

5.3.1.2. *Lower bounds on* λ

The useful bound

$$\lambda > \lambda_M^{\frac{M}{M+1}}$$

has been proved [30]. The above evaluation of λ_2, see Eq. (5.9), immediately yields $\lambda > 1.4892....$

Based on exact enumeration, in Ref. [29] the exact generating functions $G_M(z) = \sum_L C_{L,M} z^L$ for $M \leq 6$ are given. For $M = 3$ they find:

$$G_3(z) = \frac{[1, -4, -4, 36, -39, -26, 50, 6, -15, 1]}{[1, -12, 54, -124, 133, 16, -175, 94, 69, -40, -12, 4, 1]},$$

where we denote by $[a_0, a_1, \ldots, a_n]$ the polynomial $a_0 + a_1 z + \cdots + a_n z^n$. As explained above, all the generating functions $G_M(z)$ are rational. For $M = 4, 5, 6$, their numerator and denominators were found to have degree $(26, 27), (71, 75)$ and $(186, 186)$ respectively, in an obvious notation. From these follow the bounds: $\lambda_3 = 1.76331...$, $\lambda_4 = 1.75146...$, $\lambda_5 = 1.74875...$ and $\lambda_6 = 1.74728...$ from which follows $\lambda > 1.61339....$

However, a stronger lower bound can be obtained from a different class of SAW, called a transverse SAW [29]. These are just SAW that go from the left boundary to the right boundary of a $L \times L$ square, and are therefore a superset of the SAW crossing a square under consideration.

If T_L denotes the number of such transverse SAW on the $L \times L$ lattice, then it has been proved [29] that

$$\lambda \geq T(L)^{1/((L+1)(L+2))}. \tag{5.11}$$

From enumerations of $T(L)$, the improved bound $\lambda > 1.6284$ was obtained [29]. Combining these results for lower and upper bounds finally gives

$$1.6284 < \lambda < 1.781684.$$

5.3.2. Results

As has been discussed [29], to obtain the exact value of the number of SAW crossing a square, some of which are integers with nearly 100 digits, the enumerations were performed several times, each time *modulo* a different small prime. This saves storage space, and the cost of multi-precision calculation. The enumerations were then reconstructed using the Chinese Remainder Theorem. Each run for a 19×19 lattice took about 72 hours using 8 processors of a multiprocessor 1 GHz Compaq Alpha computer. Ten such runs were needed to uniquely specify the resultant numbers.

Proceeding as above, Bousquet-Mélou *et al.* [29] calculated $c_n(L)$ for all n for $L \leq 17$. In addition, they computed $C_{18}(1)$ and $C_{19}(1)$, the total number of SAW crossing an 18×18 and 19×19 square respectively. These are given in Table 5.3.2.

5.3.3. Numerical analysis

It has been proved [30, 31] that $\lim_{L \to \infty} C_{L,L}^{\frac{1}{L^2}} = \lambda$ exists. From this it is likely that $R_L = C_{L+1,L+1}/C_{L,L} \sim \lambda^{2L}$ though this has not been proved. Accepting this, the generating function $\mathcal{R}(x) = \sum_L R_L x^L$ will have radius of convergence $x_c = 1/\lambda^2$, which can be estimated accurately using differential approximants [34]. In this way it was estimated that for the crossing problem $x_c = 0.32858(5)$. An alternative determination gave the even more precise value $x_c = 0.328574(2)$. Hence $\lambda = 1.744550(5)$, which is the most precise numerical estimate of λ available.

Whittington and Guttmann [31] and later Burkhardt and Guim [35] also studied the behaviour of the mean number of steps in a path on an $L \times L$ lattice

$$\langle n(x, L) \rangle = \frac{\sum_n n c_n(L) x^n}{\sum_n c_n(L) x^n} \tag{5.12}$$

Table 1. The total number of walks crossing a square, $C_L(1)$.

L	$C_L(1)$
1	2
2	12
3	184
4	8512
5	1262816
6	575780564
7	789360053252
8	3266598486981642
9	41044208702632496804
10	1568758030464750013214100
11	182413291514248049241470885236
12	64528039343270018963357185158482118
13	6945066476152136166427470154890735899 6488
14	22744971467681273963182645932798986338 7613323440
15	22667455688626727463745673967130989348 66324885408319028
16	68745445609149931587631563132489232824 58794596809945728 5419306
17	63448146112379639713102975407955244004 9494439868664806936463 69387855336
18	17821128408420651298933849466523252751 6783806570476765593145247 4605826 692782532
19	15233449717048799930807428103192296908 9945425532329455577602986673 7355 060592877569255844

as well as the fluctuations of this quantity

$$V(x, L) = \frac{\sum_n n^2 c_n(L) x^n}{\sum_n c_n(L) x^n} - \langle n(x, L) \rangle^2 \qquad (5.13)$$

which is a kind of heat capacity. As discussed above, a phase transition takes place as one varies the fugacity x associated with the walk length. At a critical value, x_c the average walk length of a path on an $L \times L$ lattice changes from $\Theta(L)$ to $\Theta(L^2)$.

As noted before [31], the critical fugacity was estimated, and conjectured to be $x_c = 1/\mu$, [31], with the conjecture proved in Ref. [32]. In Ref. [29] the behaviour at $x = x_c$ was also studied, and it was found that $\langle n(x, L) \rangle = \Theta(L^{1/\nu})$ where the numerical evidence is consistent with $\nu = 3/4$. Similar conclusions were drawn earlier [35]. For any given value of L the fluctuation $V(x, L)$ is observed to have a single maximum located at $x_c(L)$. The behaviour of $V(x, L)$ was studied in detail also [29]. It was found to obey a standard finite-size scaling ansatz,

$$V(x, L) \sim L^{2/\nu} \tilde{V}((x - x_c) L^{1/\nu}), \qquad (5.14)$$

where $\tilde{V}(y)$ is a scaling function. From this it follows that the position and the height of the peak in $V(x, L)$ scale as $x_c(L) - x_c \sim L^{-1/\nu}$ and $V_{\max}(L) \sim L^{2/\nu}$ respectively.

The data for the mean-length at x_c, and the position and height maxima of the fluctuations were analysed by forming the associated generating functions, $N(z) = \sum_L \langle n(x, L) \rangle z^L$ etc., and using differential approximants to analyse the generating function. Given the expected asymptotic behaviour of these quantities, the generating functions are expected to have a singularity at $z_c = 1$ with critical exponents $-1/\nu - 1$ (average length at x_c), $1/\nu - 1$ (position of the peak), and $-2/\nu - 1$ (height of the peak). The observed behaviour was fully consistent with these expectations, and with an exponent value $\nu = 3/4$.

Many other properties of this model have been specified [29], but the above discussion covers the most interesting aspects from the point of view of a model of phase transitions.

5.4. Discussion

In this article we have tried to show the richness of SAW models as canonical models of phase transitions which possess both pedagogical value, and are capable of modelling a wide variety of physically interesting models. In the introduction we first discussed the most common type of interactions, monomer-monomer interactions, then monomer-wall interactions and finally a combination of both.

In the next section we showed how these ideas, and others, could be applied to a number of important biological situations, and studied three such problems in more detail. Finally, in the last section, we discussed in rather more detail the model of SAW crossing an $L \times L$ square. This is perhaps the most transparent and persuasive model of a phase transition, where the physical mechanism is entirely clear.

Acknowledgments

In choosing this topic for this memorial volume, I have in mind Ken Hines's qualities as a great teacher and a wonderful physicist, as well as an enthusiast for his profession. It was Ken who both welcomed me into the world of physics, and facilitated that entry; actions for which I am forever grateful.

I wish to thank Andrew Rechnitzer, who provided most of the figures in this paper, and my colleagues Mireille Bousquet-Mélou, Iwan Jensen and Christoph Richard for permission to quote extensively from aspects of our joint work. This work was supported by the Australian Research Council, to whom I express my gratitude.

References

[1] N. Clisby, *Phys. Rev. Lett.* (submitted) (2009).

[2] T. Prellberg and A. L. Owczarek, *Phys. Rev. E* **62**, 3780 (2000).

[3] J. Krawcyzk, A. L. Owczarek, T. Prellberg and A. Rechnitzer, *Europhys. Lett.* **70**, 1 (2005).

[4] R. M. Wartell and A. S. Benight, *Phys. Rep.* **126**, 67 (1985).

[5] D. Poland and H. A. Scheraga, *J. Chem. Phys.* **45**, 1456 (1966).

[6] D. Poland and H. A. Scheraga, *J. Chem. Phys.* **45**, 1464 (1966).

[7] M. E. Fisher, *J. Chem. Phys.* **45**, 1469 (1966).

[8] S. Leibler, R. R. P. Singh, and M. E. Fisher, *Phys. Rev. Letts.* **59**, 1989 (1987).

[9] M. E. Fisher, A. J. Guttmann and S. G. Whittington, *J. Phys. A: Math. Gen.* **24**, 3095 (1991).

[10] J. A. Banavar, A. Maritan, and A. Stella, *Phys. Rev. A* **43** , 5752 (1991).

[11] K. Svoboda and S. M. Block, *Ann. Rev. Biophys. Biomol. Struct.* **23**, 247 (1994).

[12] A. Ashkin, *Proc. Natl. Acad. Sci. USA* **94**, 4853 (1997).

[13] M. S. Causo, B. Coluzzi, and P. Grassberger, *Phys. Rev. E* **62**, 3958 (2000).

[14] E. Carlon, E. Orlandini, and A. L. Stella, *Phys. Rev. Lett.* **88**, 198101 (2002).

[15] M. Baiesi, E. Carlon, and A. L. Stella, *Phys. Rev. E* **66**, 21804 (2002).

[16] M. Baiesi, E. Carlon, Y. Kafri, D. Mukamel, E. Orlandini, and A. L. Stella, *Phys. Rev. E* **67**, 21911 (2002).

[17] Y. Kafri, D. Mukamel, and L. Peliti, *Phys. Rev. Lett.* **85**, 4988 (2000).

[18] Y. Kafri, D. Mukamel, and L. Peliti, *Eur. Phys. J. B* **27**, 135 (2001).

[19] Y. Kafri, D. Mukamel, and L. Peliti, *Physica A* **306**, 39 (2002).

[20] C. Richard and A. J. Guttmann, *J. Stat. Phys.* **115**, 925 (2004).

[21] D. C. Brydges and J. Z. Imbrie, *Ann. Math.* **158**, 1019 (2003).

[22] C. Richard A. J. Guttmann, and I. Jensen, *J. Phys. A: Math. Gen.* **34**, L495 (2001).

[23] B. Nienhuis, *Phys. Rev. Lett.* **49**, 1062 (1982).

[24] C. Richard, *J. Stat. Phys.* **108**, 459 (2002).

[25] E. Orlandini, M. Tesi, and S. G. Whittington, *J. Phys. A: Math. Gen.* **34**, 1535 (2005).

[26] P. K. Mishra, S. Kumar, and Y. Singh, *Europhys. Letts.* **69**, 102 (2005).

[27] S. Kumar, I. Jensen, J. L. Jacobsen, and A. J. Guttmann, *Phys. Rev. Lett.* **98**, 128101 (2007)

[28] N. Madras and G. Slade, *The Self-Avoiding Walk* (Boston: Birkhäuser) (1993).

[29] M. Bousquet-Mélou, A. J. Guttmann, and I Jensen, *J. Phys. A:Math. Gen.* **38**, 9159 (2005).

[30] H. L. Abbott and D. Hanson, *Ars Combinatoria* **6**, 163 (1978).

[31] S. G. Whittington and A. J. Guttmann, *J. Phys A:Math. Gen,* **23**, 5601 (1990).

[32] N. Madras, *J. Phys. A: Math. Gen.* **28**, 1535 (1995).

[33] R. P. Stanley in *Enumerative combinatorics. Vol. 2*, volume 62 of *Cambridge*

Studies in Advanced Mathematics, Cambridge University Press, Cambridge, (1999).

[34] A. J. Guttmann, Asymptotic analysis of power-series expansions in *Phase Transitions and Critical Phenomena vol. 13* (eds. C Domb and J L Lebowitz) (New York: Academic), (1989).

[35] T. W. Burkhardt and I. Guim, *J. Phys. A: Math. Gen.* **24**, L1221 (1991).

Chapter 6

Aspects of Plasma Physics

Roger J. Hosking

Faculty of Science, UBD, Gadong BE1410, Brunei

Email: marjhg@gmail.com

6.1. Preamble

A year or so after I took up my first academic job in early 1966, at what
became The Flinders University of South Australia, following my doctorate
in Canada and a post-doctoral at the Max Planck Institute in Munich,
Graeme Lister became my first PhD student on the recommendation of Ken
Hines. My first wife and I shared a table at the reception when Graeme
was married shortly afterward, and Ken and Suza Hines instantly became
two of our closest personal friends. Ken's charm and mana were more than
enough to persuade the waitress at the reception that we deserved extra
helpings and rather more wine at our table – and on many of my later
visits to Melbourne I would take every opportunity to join Ken and the
other regulars at the Thursday lunch club in Lygon Street or at a nearby
pub! Our families also met on many occasions over the years, either at
the Hines' North Balwyn family home or at one of their two beach houses
at "Moggs" near Lorne. (The first beach house was destroyed by bushfire,
leaving only melted glass from a few essential provisions kept there!)

There were also the annual AINSE (Australian Institute of Nuclear Science and Engineering) plasma physics meetings at Lucas Heights (Sydney), which I first attended in early 1967. On one occasion, Ken gave a
memorable after dinner speech in which he referred to the Father (Charles
Watson-Munro at Sydney), the Son (Max Brennan, then at Flinders) and
the Holy Ghost of Plasma Physics (at Melbourne). On another occasion,
Ken and I were able to tempt Bob Dewar to abandon one formal session (on
E/p, it must be said) to brave quite dangerous surf at the nearby suburb

of Cronulla. Ken's mates at Lucas Heights would also usually ensure that we enjoyed at least one good lunch during these visits to Sydney, such as at Doyles at Watsons Bay.

One exception was after the 1972 meeting when I had the bus driver divert to Sydney Airport, since I had to take an early flight back to New Zealand – when Ernest Titterton, also on the bus, observed that of course I had to rush back to rejoin my "boss". The reference was to Bruce Liley, whose departure from the Australian National University (ANU) was largely precipitated by this successor to the much-loved Mark Oliphant. Bruce had taken the Chair of Physics at the University of Waikato; and after working with Bruce for a few months at ANU in 1968/9, I had taken up an appointment in Mathematics at the same University in early 1971.

6.2. Introduction

Bruce Liley derived an invariant form for the plasma pressure tensor [1], which reduces in Cartesian coordinates to the form given in the famous monograph by Chapman and Cowling [2]. I took Bruce's result with me during my study leave at Culham Laboratory in 1970, and found that a widely applicable expansion for magnetoplasma yielded parallel viscosity, transverse (or finite Larmor radius) and perpendicular viscosity contributions as successive terms. This was reported in an otherwise deliberately conservative paper in *Plasma Physics* which addressed Cartesian geometry [3]. The implied dominance of the leading parallel viscosity contribution, including the magnetic curvature term in curvilinear coordinates, was nevertheless recognised at the time and in the conclusion to that paper. Ray Grimm (my companion on the bus each working day between Harwell site and Culham) combined with John Johnson on another contemporary paper [4], in which the magnetic field curvature was considered. Subsequent work with Graeme Lister and Derek Robinson showed that parallel viscosity stabilises "resistive-g" modes, including the interchange modes of major interest in thermonuclear fusion devices. However, papers on plasmas in which a classical shear viscosity is used continue to appear, despite the fact that this form corresponds to a zero magnetic field in Bruce's result, and at best, is a poor replica of the third order perpendicular viscosity term in the expansion! Ken Hines was aware of this work from AINSE presentations and our conversations. Indeed, Ken was ever conscious that abuse of power by some can be to the detriment of scientific endeavour, and he was especially interested and supportive. It is therefore fitting to dedicate the discussion below to Ken, as part of the workshop memorial to him held at Melbourne on 18 March 2005, and also to my other late friend and colleague Bruce Liley.

6.3. Boltzmann equation

On a microscopic scale, the motion of gases and liquids (fluids) is defined by the dynamics of their constituent atoms or molecules – and in the case of ionised gases (plasmas), their constituent ions and electrons. However, in fluid mechanics and magnetohydrodynamics (MHD) the motion of fluids and plasmas is described on a macroscopic scale using an appropriate complete mathematical model, typically including equations which correspond to the microscopic fluid or plasma constituents being correlated in some way.

Macroscopic equations may be derived from an integro-differential equation, first established by Boltzmann in 1872 to describe the classical dynamics of gases. Thus for fluids it is normally assumed that the atoms or molecules are correlated due to binary collisions, over distances much less than the macroscopic length scale and at frequencies much greater than any characteristic macroscopic frequency. The extension of this assumption from fluids to plasmas may be justified because the electric field of any charge is screened out on a scale characterised by the Debye length, so that charge (ion and electron) collisions are predominantly binary over distances bounded below by the Coulomb length and above by the Debye length.

For each constituent species s, the Boltzmann equation defines the evolution in time t of a velocity distribution function, $f_s(\mathbf{r}, \mathbf{c}, t)$, in phase space (position $\mathbf{r}$, velocity $\mathbf{c}$) such that the number density of that species is

$$n_s(\mathbf{r}, t) = \int f_s(\mathbf{r}, \mathbf{c}, t) \, d\mathbf{c} \, . \tag{6.1}$$

Thus, omitting the species subscript s for convenience, the governing equation for each distribution function is

$$\frac{\partial f}{\partial t} + \mathbf{c} \cdot \boldsymbol{\nabla} f + \mathbf{a}(\mathbf{r}, \mathbf{c}, t) \cdot \boldsymbol{\nabla}_{\mathbf{c}} f = \frac{\delta f}{\delta t} \, , \tag{6.2}$$

where $\mathbf{a}$ is the acceleration of that species (molecule or ion) and $\boldsymbol{\nabla}_{\mathbf{c}}$ is the gradient operator relative to the independent velocity vector (analogous to $\boldsymbol{\nabla}$ relative to the position vector) in phase space. For a species of mass m and charge e, the acceleration,

$$\mathbf{a}(\mathbf{r}, \mathbf{c}, t) = \mathbf{g} + \frac{e}{m} \left[\mathbf{E}(\mathbf{r}, t) + \mathbf{c} \times \mathbf{B}(\mathbf{r}, t) \right],$$

includes both gravitational and electromagnetic components (**g**, **E** and **B** denote the respective gravitational, electric and magnetic fields). The right-hand side of Eq. (6.2) represents the time rate of change of the distribution function due to the microscopic collisions.

6.4. Moment equations

Fundamental macroscopic quantities in fluid mechanics and MHD are related to low level moments of the velocity distribution function. Thus if $\mathbf{w} = \mathbf{c} - \mathbf{c_0}$ denotes the peculiar velocity relative to some reference velocity $\mathbf{c_0}$, typically a macroscopic fluid velocity (specified later), the moment corresponding to any tensor quantity $\mathcal{F}(\mathbf{w})$ is defined by

$$n < \mathcal{F} > = \int f(\mathbf{r}, \mathbf{c}, t) \, \mathcal{F}(\mathbf{w}) \, d\mathbf{c} . \qquad (6.3)$$

For each species, the low level moments of major interest are then

$$
\begin{aligned}
\rho &\equiv n < m > & &\text{(density)} \\
\mathbf{u} &\equiv <\mathbf{w}> & &\text{(mean relative velocity)} \\
p &\equiv n < \tfrac{1}{3}mw^2 > & &\text{(hydrostatic pressure)} \\
\mathcal{P} &\equiv n < m\mathbf{w}\mathbf{w} > & &\text{(pressure tensor)} \\
\mathbf{q} &\equiv n < \tfrac{1}{2}mw^2\mathbf{w} > & &\text{(thermal flux)}
\end{aligned}
$$

where n is the particle number density and $\mathcal{P} = p\,\mathcal{I} + \mathcal{T}$ has the traceless part $\mathcal{T} = n <m\{\mathbf{w}\mathbf{w}\}>$ to be discussed in Sec. 6.5.[a]

A general equation of change for any moment, Eq. (6.3), may be derived from the Boltzmann equation, Eq. (6.2). This simplification of Enskog's equation (cf. Chapman and Cowling [2]), the basic form adopted here, is

$$
\begin{aligned}
\mathbf{I}(\mathcal{F}) = &\frac{dn < \mathcal{F} >}{dt} + n < \mathcal{F} > \boldsymbol{\nabla}\cdot\mathbf{c_0} + \boldsymbol{\nabla}\cdot(n < \mathcal{F}\mathbf{w} >) \\
&- n\left[< \frac{d\mathcal{F}}{dt} > + < \mathbf{w}\cdot\boldsymbol{\nabla}\mathcal{F} > + \mathbf{f}\cdot < \boldsymbol{\nabla}_\mathbf{w}\mathcal{F} > \right. \\
&\left. + \frac{e}{m} < \mathbf{w}\times\mathbf{B}\cdot\boldsymbol{\nabla}_\mathbf{w}\mathcal{F} > - < (\boldsymbol{\nabla}_\mathbf{w}\mathcal{F})\mathbf{w} > :\boldsymbol{\nabla}\mathbf{c_0} \right], \quad (6.4)
\end{aligned}
$$

[a]Note: $\mathcal{T} \equiv \{\mathcal{P}\}$, where the operator $\{\ \}$ gives a symmetric traceless result; *viz.*

$$\{\mathcal{P}\} \equiv \frac{1}{2}(\mathcal{P} + \mathcal{P}^T) - \frac{1}{3}\mathcal{P}{:}\mathcal{I}\,\mathcal{I} ,$$

The superscript T denotes the transpose and $\mathcal{I}$ is the unit dyadic.

involving the time derivative,

$$\frac{d}{dt} = \frac{\partial}{\partial t} + \mathbf{c_0} \cdot \nabla ,$$

the acceleration,

$$\mathbf{f} = \mathbf{g} + \frac{e}{m} [\mathbf{E} + \mathbf{c_0} \times \mathbf{B}] - \frac{d\mathbf{c_0}}{dt} ,$$

and the collision integral,

$$\mathbf{I}(\mathcal{F}) \equiv \int \frac{\delta f}{\delta t} \mathcal{F} \, d\mathbf{c} . \tag{6.5}$$

If $\mathcal{F}$ is successively identified with m, $m\mathbf{w}$, $\frac{1}{2}mw^2$, $m\{\mathbf{ww}\}$, $\frac{1}{2}mw^2\mathbf{w}$...,
Eq. (6.4) generates the macroscopic equations of fluid mechanics or MHD
(namely conservation of mass, momentum, kinetic energy, and then much
less familiar forms). However, this hierarchy of moment equations is not
closed because a higher moment is incorporated at each level.

Closure at some low level, implicit in any fluid mechanics or MHD
model, may be achieved by approximating the distribution function in some
acceptable way. For example, the distribution function may be expanded in
multi-dimensional Hermite polynomials to provide a "thirteen moment approximation", an approach due to Grad[5]. Thus the corresponding closure
of the moment equations occurs at a level involving no more than thirteen
scalar fields, namely, the moments ρ, $\mathbf{u}$, p, $\mathcal{T}$, and $\mathbf{q}$. Collisions tend to
drive the distribution function toward the Maxwell velocity distribution,

$$f^0 = n \left(\frac{\alpha}{\pi}\right)^{\frac{3}{2}} \exp\left(-\alpha w^2\right) , \tag{6.6}$$

which is traditionally said to characterise "local thermodynamic equilibrium", when it is usually considered that the peculiar velocity $\mathbf{w}$ is isotropic.
However, a truncated expansion could obviously be taken about any zeroth
order distribution function, and would also lead to a system of macroscopic
equations such as outlined in the following Section, but with associated
modifications to the moments defined above.

6.5. Macroscopic equations

Assuming that contributions from the three lowest level integrals
$I(m), I(m\mathbf{w})$ and $I(mw^2/2)$ vanish, i.e. that the binary elastic collisions
over all species preserve mass, momentum and kinetic energy, the corresponding macroscopic equations of mass conservation (continuity), motion,

and (kinetic) energy are respectively,

$$\frac{d\rho}{dt} + \rho\boldsymbol{\nabla}\cdot\mathbf{v} = 0, \tag{6.7}$$

$$\rho\frac{d\mathbf{v}}{dt} + \boldsymbol{\nabla}\cdot\mathcal{P} = \mathbf{j}\times\mathbf{B} + \rho\mathbf{g}, \tag{6.8}$$

and

$$\frac{d}{dt}\left(\frac{3}{2}p\right) + \frac{3}{2}p\boldsymbol{\nabla}\cdot\mathbf{v} + \mathcal{P}{:}\boldsymbol{\nabla}\mathbf{v} + \boldsymbol{\nabla}\cdot\mathbf{q} = \mathbf{j}\cdot(\mathbf{E} + \mathbf{v}\times\mathbf{B}). \tag{6.9}$$

The total mass, momentum and current densities are the species sums

$$\rho \equiv \Sigma_s\rho_s , \quad \rho\mathbf{v} \equiv \Sigma_s\rho_s <\mathbf{c}_s> , \quad \mathbf{j} \equiv \Sigma_s\frac{e_s}{m_s}\rho_s <\mathbf{c}_s> ,$$

and the time derivative,

$$\frac{d}{dt} = \frac{\partial}{\partial t} + \mathbf{v}\cdot\boldsymbol{\nabla},$$

is often referred to as the material derivative since the reference velocity $\mathbf{c_0}$ is identified here with the mass-weighted mean velocity $\mathbf{v}$. (The double dot product term on the left-hand side of Eq. (6.9) is of course a scalar, since, if $\mathcal{A}$ and $\mathcal{B}$ are any two dyadics, $\mathcal{A}{:}\mathcal{B} = \mathcal{A}_{ij}\mathcal{B}_{ji}$.)

Although an electric force term does not appear in the equation of motion, Eq. (6.8), because a plasma is assumed to be quasi-neutral on the macroscopic scale, having the gravitational force $\rho\mathbf{g}$ supplemented with the electromagnetic body force $\mathbf{j}\times\mathbf{B}$ is of major interest in MHD (i.e. whenever the medium is electrically conducting). From Eq. (6.7), note also that the energy equation, Eq. (6.9), can be rewritten

$$\frac{3}{2}\rho^{\frac{5}{3}}\frac{d}{dt}\left(p\rho^{-\frac{5}{3}}\right) = -\mathcal{T}{:}\boldsymbol{\nabla}\mathbf{v} - \boldsymbol{\nabla}\cdot\mathbf{q} + \mathbf{j}\cdot(\mathbf{E} + \mathbf{v}\times\mathbf{B}). \tag{6.10}$$

Thus, identifying $\gamma = \frac{5}{3}$ in the well known adiabatic equation of state,

$$\frac{d}{dt}(p\rho^{-\gamma}) = 0 \quad \text{or} \quad \frac{dp}{dt} + \gamma p\boldsymbol{\nabla}\cdot\mathbf{v} = 0, \tag{6.11}$$

corresponds to negligible energy dissipation and thermal conduction, and (in a conducting medium) negligible electromagnetic heating.

In summary, given a barotropic equation of state such as Eq. (6.11), the macroscopic motion is defined via Eqs. (6.7) and (6.8), provided a suitable constitutive relation for the pressure tensor $\mathcal{P}$ can be invoked. Thus in fluid mechanics there is usually a complete (closed) system of five scalar equations in the field variables $\{\rho, \mathbf{v}, p\}$ defining the dynamics, as for example, when the simple form $\mathcal{T} = -2\mu\{\boldsymbol{\nabla}\mathbf{v}\}$ is adopted for the traceless, 'non-hydrostatic', component (see below). In the case of MHD, where the electromagnetic body force $\mathbf{j}\times\mathbf{B}$ is significant, in addition to an appropriate form for the pressure tensor, supplementary electromagnetic equations need to be included as well. On the other hand, a constitutive relation for the thermal flux vector $\mathbf{q}$ is needed only if the thermal conduction term $\boldsymbol{\nabla}\cdot\mathbf{q}$ is significant in the energy equation.

6.6. Plasma pressure tensor

An appropriate constitutive relation for the pressure tensor $\mathcal{P} = p\mathcal{I} + \mathcal{T}$ in a plasma is now considered. Identifying $\mathcal{F} \equiv m\{\mathbf{ww}\}$ in the Enskog equation, Eq. (6.4), yields the moment equation for the traceless component $\mathcal{T}$, for each species. This moment equation implies that the classical constitutive relation of fluid mechanics should usually be modified for a plasma in a magnetic field. For a two-species (ion-electron) plasma, a suitable linear representation of the collision integral consistent with tensor rank is $\mathbf{I}(m\{\mathbf{ww}\}) = (\beta/\tau)\mathcal{T}$, where β is an appropriate coefficient and τ denotes the microscopic particle collision time, and the plasma constitutive equation is implicit in the moment equation expressed in the form,

$$\mathcal{T} - 2\{\mathcal{T}\times\mathbf{a}\} = -2\mu\mathcal{S} . \qquad (6.12)$$

That involves a generalised rate of deformation tensor $\mathcal{S}$ associated with the viscosity coefficient $\mu = p\tau/\beta$ and a magnetic field term where

$$\mathbf{a} \equiv \frac{e\mathbf{B}}{m}\frac{\tau}{\beta} .$$

However, The coefficient μ and vector $\mathbf{a}$ should reflect not only the collision integral but also the increasingly dominant inertia term as the parameter $\omega\tau$ (where ω denotes the characteristic macroscopic frequency) ranges from small "collisional" to large "collisionless" values. The corresponding explicit form for the traceless component of the plasma pressure tensor, obtained by Bruce Liley and as derived in the Appendix, is

$$\mathcal{T} = -\frac{2\mu}{(1+|\mathbf{a}|^2)\,(1+4|\mathbf{a}|^2)}\left[(\mathcal{S}+2\{\mathcal{S}\times\mathbf{a}\})(1+|\mathbf{a}|^2)\right.$$
$$\left. + 6\,(\,\{\mathcal{S}\cdot\mathbf{aa}\}+2\{\{\mathcal{S}\cdot\mathbf{aa}\}\times\mathbf{a}\}\,)+6\,\mathcal{S}{:}\mathbf{aa}\{\mathbf{aa}\}\right]. \qquad (6.13)$$

This invariant form, Eq. (6.13), reduces to the fluid mechanics relation $\mathcal{T} = -2\mu\{\boldsymbol{\nabla}\mathbf{v}\}$ when $\mathbf{a} \equiv 0$ and $\mathcal{S} \equiv \{\boldsymbol{\nabla}\mathbf{v}\}$, and yields the Cartesian representation in a magnetic field [2] when $|\mathbf{a}| \neq 0$. Moreover, it renders other representations in curvilinear coordinate systems (including magnetic field curvature terms) of significant interest in MHD.

Except in the neighbourhood of a magnetic null, in magnetised plasma (i.e. plasma with a permeating magnetic field) one typically has $|\mathbf{a}| \gg 1$. Hence, on noting

$$-\frac{2\mu}{(1+|\mathbf{a}|^2)(1+4|\mathbf{a}|^2)} = \frac{1}{2}\mu \left[-|\mathbf{a}|^{-4} + \frac{5}{4}|\mathbf{a}|^{-6} + \cdots\right],$$

one may expand Eq. (6.13) to obtain

$$\mathcal{T} = \mathcal{T}_{\parallel} + \mathcal{T}_L + \mathcal{T}_{\perp} + \cdots ,$$

where, with $\mathbf{b} \equiv \mathbf{B}/|\mathbf{B}|$,

$$\mathcal{T}_{\parallel} = -3\mu \mathcal{S}: \mathbf{bb}\{\mathbf{bb}\}, \tag{6.14}$$

$$\mathcal{T}_L = -\frac{\mu}{|\mathbf{a}|}\left\{\mathcal{S}\times\mathbf{b} + 6\{\mathcal{S}\cdot\mathbf{bb}\}\times\mathbf{b}\right\}, \tag{6.15}$$

$$\mathcal{T}_{\perp} = -\frac{\mu}{2|\mathbf{a}|^2}\left[\mathcal{S} + 6\{\mathcal{S}\cdot\mathbf{bb}\} - \frac{15}{2}\mathcal{S}: \mathbf{bb}\{\mathbf{bb}\}\right], \tag{6.16}$$

are the parallel, the transverse (or "FLR"), and perpendicular components respectively [3].

The components, $\mathcal{T}_{\parallel}$ and $\mathcal{T}_{\perp}$, are named with reference to the magnetic field direction $\mathbf{b}$, while the component $\mathcal{T}_L$ is named with reference to its "finite Larmor radius" derivation from the collisionless Vlasov equation (when it is the assumed that the only microscopic particle correlations are due to the magnetic field). Although these results apply to each plasma species, the viscosity coefficient is typically much larger for the ions than for the electrons in a two-species (ion-electron) plasma. Thus the dominant contribution is the ion viscosity, when the adoption of $\mathcal{S} = \{\boldsymbol{\nabla}\mathbf{v}\}$ familiar from fluid mechanics may be justifiable. [b] The leading parallel component, Eq. (6.14), then involves

$$\mathcal{S}:\mathbf{bb} = \mathbf{b}\cdot\boldsymbol{\nabla}(\mathbf{v}\cdot\mathbf{b}) - \mathbf{v}\cdot(\mathbf{b}\cdot\boldsymbol{\nabla}\mathbf{b}) - \frac{1}{3}\boldsymbol{\nabla}\cdot\mathbf{v} . \tag{6.17}$$

The second term on the right-hand side of Eq. (6.17) is proportional to the magnetic field curvature $\kappa \equiv \mathbf{b}\cdot\boldsymbol{\nabla}\mathbf{b}$. As previously mentioned, parallel

---

[b]When $\mathcal{S} = \{\boldsymbol{\nabla}\mathbf{v}\}$ is adopted, the successive contributions rendered by Eqs. (6.14) to (6.16) essentially agree with the results obtained in various ways by several other authors [6].
cf. also the note added in proof.

ion viscosity proved to be particularly important for resistive stability theory [3].

Note added in proof:

Alternative forms of the results given by Eqs. (6.14), (6.15), and (6.16) were obtained by J.D. Callen *et al.* (*Plasma Physics and Controlled Nuclear Research*, 157, IAEA, Vienna 1987), as has been mentioned in a recently expanded discussion by the present author entitled "Extended MHD Equations" (in *Frontiers of Plasma Physics and Technology*, IAEA, Vienna 2008), where the direct use of the fundamental moment equations obtained from Eq. (6.4) is further emphasised. Consistent alternative forms of the relevant higher moment equations have recently been presented by J. Ramos (Physics of Plasmas **12**, 052102, 2005; **14**, 052506, 2007), where however, the moments are defined with reference to the macroscopic mean flow velocity of each particular species as in Braginskii [6], rather than the universal mass-weighted mean velocity of the system as in Chapman and Cowling [2] and adopted by the present author. These alternative forms have been re-derived using symbolic tensor notation, in a presentation to appear in the Proceedings of the 4th International Conference on Plasma Physics and Technology (held at Kathmandu 6-10 April 2009), also to be published by the IAEA.

Appendix A. Derivation of Eq. (6.13)

The derivation of the explicit result for the traceless component of the plasma pressure tensor from the implicit form given in Eq. (6.12), is accomplished using certain vector and tensor identities.

If $\mathcal{T}$ is a dyadic and $\mathbf{a}$ is a vector, the cross product is

$$\mathcal{T} \times \mathbf{a} = \mathcal{T}_{ij} \, a_k \, \epsilon_{jkl} \, \mathbf{e}_i \mathbf{e}_l \,, \qquad (A.1)$$

where ϵ_{jkl} denotes the alternating tensor and $\{\mathbf{e}_i\}$ any Cartesian (self-reciprocal) basis set. The usual index notation is assumed, as is the convention that repeated indices denote summation over $1, 2, 3$.

If $\mathcal{T}$ is symmetric ($\mathcal{T}^T = \mathcal{T}$), then

$$(\mathcal{T} \times \mathbf{a})^T = -\mathbf{a} \times \mathcal{T} \qquad \text{and} \qquad \mathcal{I} : \mathcal{T} \times \mathbf{a} = 0, \qquad (A.2)$$

where $\mathcal{I} = \mathbf{e}_i \mathbf{e}_i$ is the unit dyadic. On applying the bracket operator defined in the footnote on page 4, one has,

$$2 \{\mathcal{T} \times \mathbf{a}\} = \mathcal{T} \times \mathbf{a} - \mathbf{a} \times \mathcal{T}$$

and thus, on substituting $\{\mathcal{T} \times \mathbf{a}\}$ for $\mathcal{T}$

$$4\,\{\{\mathcal{T} \times \mathbf{a}\} \times \mathbf{a}\} \;=\; (\mathcal{T} \times \mathbf{a}) \times \mathbf{a} - 2\,\mathbf{a} \times \mathcal{T} \times \mathbf{a} + \mathbf{a} \times (\mathbf{a} \times \mathcal{T}) \,. \tag{A.3}$$

Note that brackets are unnecessary in the second triple product on the right-hand side, and there are a set of relations,

$$(\mathbf{a} \times \mathcal{T}) \times \mathbf{a} \;=\; \mathbf{a} \times (\mathcal{T} \times \mathbf{a}),$$

$$2\,\{\mathcal{T} \times \mathbf{a}\}{:}\mathbf{aa} \;=\; (\mathcal{T} \times \mathbf{a} \;-\; \mathbf{a} \times \mathcal{T}){\cdot}\mathbf{a}{\cdot}\mathbf{a} \;=\; 0,$$

$$\mathbf{a} \times (\mathbf{a} \times \mathcal{T}) \;=\; \mathbf{aa}{\cdot}\mathcal{T} \;-\; |\mathbf{a}|^2 \mathcal{T}. \tag{A.4}$$

If the symmetric dyadic $\mathcal{T}$ is also traceless ($\mathcal{T}_{ii} = 0$), on writing out $\mathbf{a} \times \mathcal{T} \times \mathbf{a} = a_i \mathcal{T}_{jk} a_m \mathbf{e}_i \times \mathbf{e}_j \, \mathbf{e}_k \times \mathbf{e}_m$ and grouping terms one obtains

$$\mathbf{a} \times \mathcal{T} \times \mathbf{a} \;=\; -(\mathcal{T}{\cdot}\mathbf{aa} + \mathbf{aa}{\cdot}\mathcal{T}) \;+\; (\mathcal{I}\,\mathcal{T}{:}\mathbf{aa} + |\mathbf{a}|^2 \mathcal{T}) \,, \tag{A.5}$$

on noting that $\mathbf{aa}{:}\mathcal{I} = |\mathbf{a}|^2$. Eq. (A.3) can therefore be re-expressed as

$$2\,\{\{\mathcal{T} \times \mathbf{a}\} \times \mathbf{a}\} \;=\; 3\,\{\mathcal{T}{\cdot}\mathbf{aa}\} - 2\,|\mathbf{a}|^2 \mathcal{T} \,, \tag{A.6}$$

since $(\mathcal{T}{\cdot}\mathbf{aa})^T = \mathbf{aa}{\cdot}\mathcal{T}$ and $(\mathcal{T}{\cdot}\mathbf{aa}){:}\mathcal{I} = \mathcal{T}{:}\mathbf{aa}$.

Another useful identity is

$$6\,\{\mathcal{T}{\cdot}\mathbf{aa}\}{\cdot}\mathbf{a} = 3\,\mathcal{T}{\cdot}\mathbf{a}\,|\mathbf{a}|^2 \;+\; \mathbf{a}\mathcal{T}{:}\mathbf{aa} \,, \tag{A.7}$$

on noting that $\mathbf{a}{\cdot}\mathcal{T}{\cdot}\mathbf{a} = \mathcal{T}{:}\mathbf{aa}$.

Finally, on replacing $\mathcal{T}$ with $\{\mathcal{T} \times \mathbf{a}\}$ in Eq. (A.6), and comparing the result with a post cross product of Eq. (A.6) with the vector $\mathbf{a}$ followed by the bracket operator, one also obtains

$$\{\{\mathcal{T} \times \mathbf{a}\}{\cdot}\mathbf{aa}\} \;=\; \{\{\mathcal{T}{\cdot}\mathbf{aa}\} \times \mathbf{a}\}. \tag{A.8}$$

Let us now proceed to derive Eq. (6.13). A post cross product of Eq. (6.12) with the vector $\mathbf{a}$ followed by the bracket operator yields

$$-4\mu\{\mathcal{S} \times \mathbf{a}\} \;+\; 4\,\{\{\mathcal{T} \times \mathbf{a}\} \times \mathbf{a}\} = 2\,\{\mathcal{T} \times \mathbf{a}\} = 2\mu\mathcal{S} + \mathcal{T} \,, \tag{A.9}$$

and in Eq. (A.6) then produces

$$-4\mu\{\mathcal{S} \times \mathbf{a}\} \;+\; 6\{\mathcal{T}{\cdot}\mathbf{aa}\} \;-\; 2\mu\mathcal{S} \;=\; \left(1 + 4|\mathbf{a}|^2\right)\mathcal{T} \,. \tag{A.10}$$

Taking a post dot product of Eq. (A.10) with the vector $\mathbf{a}$ and invoking Eq. (A.7) yields,

$$\left(1 + |\mathbf{a}|^2\right)\mathcal{T}{\cdot}\mathbf{a} \;=\; -4\mu\{\mathcal{S} \times \mathbf{a}\}{\cdot}\mathbf{a} \;-\; 2\mu\,(\mathcal{S}{:}\mathbf{aa}\,\mathbf{a} + \mathcal{S}{\cdot}\mathbf{a}) \,, \tag{A.11}$$

since $\mathcal{T}{:}\mathbf{aa} = -2\mu\mathcal{S}{:}\mathbf{aa}$ follows from Eq. (6.12). Then, elimination of $\mathcal{T}{\cdot}\mathbf{a}$ between Eqs. (A.10) and (A.11) with the help of Eq. (A.8) produces the result given by Eq. (6.13).

References

[1] B. S. Liley, *University of Waikato Physics Report* **103** (1971).

[2] S. Chapman and T. G. Cowling, *The Mathematical Theory of Non-Uniform Gases* (Cambridge University Press, 1970).

[3] R. J. Hosking and G. M. Marinoff, *Plasma Physics* **15**, 327 (1973). Subsequent publications on the parallel viscous stabilisation of "resistive-g" modes are:

R. J. Hosking and G. G. Lister, *Plasma Physics* **15**, 931 (1973)

R. J. Hosking and D. C. Robinson, *Proceedings of the Ninth European Conference on Controlled Fusion and Plasma Physics*, Oxford, 61 (1979)

Further references on the magneto-viscous stabilisation of resistive instabilities include:

G. M. Marinoff, Ph. D. thesis, The Flinders University of South Australia (1973);

R. J. Hosking and J. Tendys *J. Comp. Phys,* **66**, 274 (1986)).

[4] R. C. Grimm and J. L. Johnson, *Plasma Physics* **14**, 617 (1972).

[5] H. Grad, *Comm. in Pure and Appl. Math.* **2**, 331 (1949).

[6] S. I. Braginskii, in *Reviews of Plasma Physics*, ed. M. A. Leontovich, **Vol. 1**, 205 (Consultants Bureau, New York, 1965)

also see T. E. Stringer and J. Connor, *Physics of Fluids* **14**, 2177 (1971).

Chapter 7

Further Properties of a Magnetised Maxwellian Plasma

Victor Kowalenko

School of Physics, University of Melbourne,
Victoria 3010, Australia
E-mail: vkowa@unimelb.edu.au

Recent developments for determining the plasma oscillations in a magnetised Maxwellian plasma are employed in calculating other properties that are derived from the longitudinal dielectric response function. First, the static screening potential is evaluated. Surprisingly this yields the standard Debye potential of a field-free plasma with no corrections for the magnetic field irrespective of the magnitude of the field. Then the dispersion relation for the ion-acoustic mode is evaluated with the corrections obtained in the weak field limit.

7.1. Introduction

I can still recall my second meeting with Dr Ken Hines. It occurred in early February, 1977 when as a 20 year old student, I appeared in his office to discuss my Honours year project. His office was quite inhospitable due to the vast number of journals, reports and books residing on the shelves that competed for space with prospective visitors. Even a huge poster of Beethoven struggled to find a niche, having to settle for life pinned to the back of the door. So, I was forced to stand while Ken, seated at his stockpiled desk, discussed what would be required for the project.

He explained to me in his deep bass voice that to be inducted into the plasma theory group, I needed to work through the first four chapters of Ichimaru's book [1] and in particular, derive the dielectric tensor given by Eq. (3.72) for the magnetised Maxwellian or classical plasma. At the time the plasma theory group consisted of Ken's closest colleague, Dr. Norm Frankel and two Ph.D. students, Ken's eldest son David, who these days is a psychiatrist, and Dr. Angelo E. Delsante, who is at CSIRO Manufacturing and Infrastructure Technology. Ken expected that I would complete the task within a month just before lectures commenced and then we would discuss the actual project.

With little understanding of what plasma physics was about, I went away with his copy of Ichimaru's book, thinking that I would now gain a real appreciation for the subject. However, once I opened the book, aside from the first chapter, I was completely lost and began to panic as to whether I would be able to carry out the exercise within the required amount of time. I can even remember approaching the more able members of my class for their advice. So, it was only with a little re-assurance from my friends and much effort that I was able to complete the task within the deadline. A great sense of relief came over me as I had passed my first real test in plasma physics.

Now that I was 'sufficiently equipped' with the basic principles of classical plasma physics, I returned to Ken to discuss my Honours project. He stated that since I was familiar with the notion of Debye screening in a field-free Maxwellian plasma and had derived the dielectric tensor for a magnetised system, my project was to calculate the static screening potential of a test-particle in a magnetised Maxwellian plasma. Whilst seeming innocuous, this problem has been solved only recently as a result of important developments in asymptotics over the past two decades. An account of these developments with the solution to the problem follows.

7.2. Screening potential of a test-particle

The screening potential at a position r from a test-particle of charge Q immersed in a plasma is given by

$$V(r) = \frac{1}{(2\pi)^3} \int d\mathbf{q} \, \frac{4\pi Q \, \exp{(i\mathbf{q} \cdot r)}}{q^2 \, \epsilon_L(\mathbf{q}, \omega)} \, , \qquad (7.1)$$

where $\epsilon_L(\mathbf{q}, \omega)$ is known as the longitudinal dielectric response function and is derived from the dielectric tensor. In the case of a mobile test-particle, which is called dynamic screening, the frequency ω equals the dot-product of the wave-vector $\mathbf{q}$ with the velocity, $\mathbf{v_0}$, of the particle, i.e. $\omega = \mathbf{q} \cdot \mathbf{v_0}$. For a static test-particle, $\omega = 0$.

From Eq. (7.1) it can be seen that the longitudinal dielectric response function must contain the physics of the particular system under study. For a field-free or unmagnetised classical plasma with number density n and temperature T, the longitudinal dielectric response function in the static limit is given by

$$\epsilon_L(\mathbf{q}, 0) = 1 + \frac{q^2}{q_D^2} , \qquad q_D = \sqrt{\frac{4\pi n e^2}{kT}} . \qquad (7.2)$$

In the above q_D is called the Debye wavenumber, while we shall express T in units of energy, thereby setting Boltzmann's constant k equal to unity. Inserting Eq. (7.2) result into Eq. (7.1) yields the famous Debye screening potential, $V(r) = Q\exp(-q_D r)/r$. The situation, however, becomes more complicated when an external magnetic field, **B**, is introduced in the z-direction of the system. The system is no longer isotropic and the longitudinal dielectric response function now depends upon the parallel (q_z) and perpendicular ($q_\perp$) components of the wave-vector. In Ch. 4 of Ref. [1] the longitudinal dielectric response function is given by

$$\epsilon_L(\mathbf{q}, \omega) = 1 + \frac{q_D^2}{q^2} \left\{ 1 + \sum_{n=0}^{\infty} \frac{\omega}{\omega - n\Omega} \left[W\left(\frac{\omega - n\Omega}{|q_z|\,(T/m)^{1/2}} \right) - 1 \right] \Lambda_n(\beta) \right\},$$

$$(7.3)$$

where $\Lambda_n(\beta) = I_n(\beta)\exp(-\beta)$ (I_n is the modified Bessel function of n-th order), $\beta = q_\perp^2 T/m\Omega^2$, $q^2 = q_z^2 + q_\perp^2$, and the cyclotron frequency, Ω, is equal to eB/mc in cgs units. The W-function of Ichimaru is defined as

$$W(Z) = (2\pi)^{-1/2} \int_{-\infty}^{\infty} \frac{x\,\exp\left(-x^2/2\right)}{x - Z - i\eta} \, dx , \qquad (7.4)$$

where η is an infinitesimal. This function is related to the more commonly used plasma dispersion function of Fried and Conte [2], but as we shall see, it is better to express the function in terms of the complementary error function.

So, there it is, my Honours year project returning to haunt me! Nonetheless, it appears to be formidable. In fact, the problem with the above form is simply too intractable to solve exactly, although interestingly, Eq. (7.2) can be obtained by following the method in Ref. [1], where one sets $\zeta = n\Omega$ and takes the $B \to 0$ limit. Then the summation becomes an integral, i.e.

$$\sum_{n=-\infty}^{\infty} \to \int_{-\infty}^{\infty} d\zeta/\Omega \quad \text{and} \quad \Lambda_n(\beta) \to |\Omega| \frac{\exp(-n^2\Omega^2/q_\perp^2 v_{Tj}^2)}{\sqrt{\pi}\,q_\perp v_{Tj}}.$$

Here v_{Tj} $(= \sqrt{2T_j/m_j})$ represents the thermal velocity of the ionic species j. Therefore, although the static screening potential could be obtained in the field-free limit, the aim of the project was to determine the correction terms for weak and strong magnetic fields.

Ken's idea was to search the literature for material on the screening potential of a mobile test-particle in a field-free Maxwellian plasma since the latter is considerably more complex than static or Debye screening. The idea was that there might be mathematical methods from the attempts at solving the dynamic screening problem for a field-free Maxwellian plasma, which could be employed to obtain forms for the static screening potential for the weak and strong magnetic field cases. As a consequence, a review of these attempts was to appear in the final report. Unfortunately, there were two major problems with this idea:

(1) there was very little material on the dynamic screening problem,
(2) what little material existed, was of no benefit for solving the magnetised case.

Therefore, the only option was to find a completely different approach. But no plasma physics text-book to this day has ever been able to study the plasma behaviour in a weak magnetic field, while the strong magnetic field behaviour has been limited to finding Bernstein resonances [3] and thus, has no connection with the screening potential. To make matters worse, by the time this became apparent, the due date for the final report was imminent. Thus, an alternative project could not be considered. The situation became even more bleak for me when the 'stars' of the class, with beaming smiles, declared that their projects were going to be published in physics journals.

Still, Ken must have seen something in my report to support me, and whilst my project was hardly a resounding success, I managed to obtain a sufficiently high mark to undertake a Ph.D. Nevertheless, my confidence in classical plasma physics had taken a beating and I was only too willing to enter the world of quantum plasma physics, particularly as the other members of the group were carrying out interesting work on the relativistic electron [4] and charged Bose gases [5]. For someone languishing in classical plasma physics, a change of topic was the ideal remedy. Ken, however, was an established classical plasma physicist, and so I was very surprised at how he embraced the new field so enthusiastically. Over the next few years until the completion of my Ph.D., we became quantum plasma physicists specialising in the behaviour of particle-anti-particle or pair plasmas.

After completing my Ph.D., I returned to classical plasma physics by becoming a research scientist in the DSTO, where my role was to develop further the magnetohydrodynamical codes modelling plasma armatures in railguns. At this stage I had no formal journal publications, although Norm Frankel had succeeded in getting my thesis accepted for publication in

Physics Reports. As time elapsed, my employers became nervous as to the status of this publication and when it appeared that little progress had been made in the two years after my departure from the School of Physics, Ken intervened by suggesting that he and I produce the final manuscript [6].

7.3. Relativistic charged boson pair plasma

At the time I told Ken that I was finding it extremely difficult to continue with the quantum pair plasma work in my spare time, although I had managed to set up the derivation of the conductivity tensor for the relativistic charged boson plasma in an external magnetic field. He then proposed that he would get his Ph.D. students involved in the work. This proposal ultimately led to two major publications.

In the first of these papers [7] Drs. Nick Witte and Ross Dawe with Ken presented a thorough treatment of the Klein-Gordon equation in a magnetic field, calculated the wave-functions and then used them to obtain the various matrix elements for the polarisation tensor, which incorporates the dielectric tensor. In the second paper [8] the derivation and an exhaustive study of the properties of the polarisation tensor for a collisionless boson pair plasma in a magnetic field was given. In addition to presenting a detailed comparison of the properties of the boson pair plasma with the corresponding fermion system, some properties at $T = 0\,\mathrm{K}$ such as the screening length and propagation of the modes parallel to the magnetic field were evaluated, but only in the infinite field limit. While the system displayed anisotropic properties similar to the magnetised Maxwellian plasma, we were unable to determine corrections to field-free and infinite field limits.

Shortly after Ref. [8] was published, Norm Frankel approached me about the calculation of the screening potential, which he claimed was contrary to his investigation of the static screening potential in the non-relativistic charged Bose gas (CBG). In conjunction with the late Dr. Steve Hore [9] he had been unable to obtain an explicit form for the screening potential, but by appealing to the system's anisotropy, they conjectured two distinct cases: (1) where the system behaved as a one-dimensional field-free CBG for modal propagation perpendicular to the magnetic field and (2) where it behaved as a two-dimensional field-free CBG for modal propagation parallel to the magnetic field. These led to particularly strange results because neither depended upon the size of the magnetic field.

The problem with obtaining explicit forms for the screening potential in a magnetised plasma, be it quantum or classical, is one of asymptotics. That is, one needs to develop asymptotic forms for the dielectric/polarisation tensor in both the weak and strong magnetic field limits. Introducing the form for the longitudinal dielectric tensor given by Eq. (7.3) into Eq. (7.1) is simply futile. A major advantage in developing such forms is that one need not confine oneself to the evaluation of the screening potential. A whole host of properties, such as the dispersion relations for longitudinal and transverse modes in the system and their damping, can be evaluated in both limits, particularly the unassailable weak magnetic field limit. Therefore, whilst the screening potential itself is not a very important problem, its solution has profound ramifications for plasma physics.

The breakthrough appeared in Ref. [10] where new asymptotic expansions for the specific Kummer function that appears in the response theory of both the relativistic [8] and non-relativistic [11] CBG were presented for strong and weak magnetic fields. Although this function is referred to as the Bose-Kummer function, it is actually a variant of the incomplete gamma function. Specifically, we developed the asymptotic expansions for

$$
S(\alpha, x) = x^{-\alpha} \, \exp\left(-i\pi\alpha\right) \, \gamma\left(\alpha, x \exp\left(i\pi\right)\right) = \sum_{n=0}^{\infty} \frac{x^n}{n! \, (\alpha + n)} \ , \qquad (7.5)
$$

where $\gamma(\alpha, x)$ represents the incomplete gamma function. The weak magnetic field case corresponds to large values of $|\alpha|$. Hence, by developing a large $|\alpha|$ expansion for $S(\alpha, x)$, in effect we were presenting a new asymptotic form for the incomplete gamma function. This special function is particularly important in asymptotics due to its appearance in the theory of terminants [12]. In deriving the new asymptotic form for weak magnetic fields, a novel graphical method was employed. This method, which was summarised in Appendix A of Ref. [13], is now known as the partition method for a power series expansion [14] and is far more powerful than the standard Taylor series approach since it is not limited to convergent power series expansions. Moreover, the new expansion was actually developed under the condition that $|\Omega/\omega_p| < 1$, where $\omega_p \, (= \sqrt{4\pi n e^2/m})$ is the plasma frequency. Not only does this include the weak magnetic field case, but it also allows for the high density case.

The asymptotic form for large magnetic fields, i.e. small $|\alpha|$, was obtained by introducing the new large $|\alpha|$ expansion into a representation of the extension $S_s(\alpha, x)$ defined by

$$
S_s(\alpha, x) = \sum_{n=0}^{\infty} \frac{x^n}{n! \, (\alpha + n)^s} = a^{-s} + \sum_{n=1}^{\infty} \frac{x^n}{n! \, n^s} \, {}_1F_0(s; ; -\alpha/n) \ . \qquad (7.6)
$$

As in the case of the large $|\alpha|$ expansion the resulting expansion possessed denominators with $\alpha + x$. For the CBG, $\alpha = \hbar q_z^2 / 2m\Omega$ and $x = \hbar q_\perp^2 / 2m\Omega$, which means that $\alpha + x = \hbar q^2 / 2m\Omega$. Hence, the denominators in the terms of both expansions were no longer anisotropic. This greatly facilitates the evaluation of the system's properties including the screening potential of a test-particle for both the weak and strong magnetic field cases. The spectacular results also appeared in Ref. [13]. In Appendix D of this work it was shown how the derivation of the large $|\alpha|$ expansion could be modified to yield the longitudinal dielectric response function of the magnetised Maxwellian plasma for $|\Omega/\omega_p| \ll 1$. In addition, by taking the limit as $\Omega \to 0$, one obtained the dispersion relation for the field-free Maxwellian plasma. It was shown in Appendix E how the famous Bernstein resonances could be obtained by introducing the small $|\alpha|$ expansion into the longitudinal dielectric response function for the system. In actual fact, it was found that the dispersion relations obtained by Bernstein are not quite correct due to the mathematical method employed in Ref. [3]. However, by using the new small $|\alpha|$ expansion, which means that $|\Omega/\omega_p| \gg 1$, the correct forms were obtained for the Bernstein resonances.

The large $|\alpha|$ expansion was particularly important as it yielded the properties of the CBG for weak magnetic fields. These properties had never been obtained before for any plasma system. Furthermore, it served as the platform for determining the dispersion relation and damping of plasma oscillations in a weakly magnetised Maxwellian plasma [15]. From an in-depth study of the asymptotics for the complementary error function in the field-free limit, it was found that Landau's result [16] for the damping of the modes in a Maxwellian plasma omitted a factor of 2, due to the fact that the Stokes phenomenon [17] had not been considered properly. Now that the behaviour of plasma oscillations in a magnetised Maxwellian plasma have been determined, the next question is whether other properties of the magnetised Maxwellian plasma can be evaluated in both the strong and weak magnetic field limits, in particular the troublesome static screening potential.

7.4. Static screening potential

To determine the static screening potential, reconsider the longitudinal dielectric function given by Eq. (7.3). Although this form is similar to other versions in plasma physics texts, we shall use the form derived by Bernstein [3]. This is

$$\epsilon_L(\mathbf{q},\omega) = 1 + \sum_j \frac{q_{Dj}^2}{q^2} - i \sum_j \frac{\omega \, q_{Dj}^2}{q^2 \, \Omega_j} \int_0^\infty dy \, \exp\left(-i\frac{\omega}{\Omega_j}y\right)$$

$$\times \exp\left(-\frac{1}{2}\mu_j y^2 \; - \; \lambda_j[1 - \cos y]\right) \;, \qquad (7.7)$$

where $\mu_j = q_z^2 v_{Tj}^2 / 2\Omega_j^2$ and $\lambda_j = q_\perp^2 v_{Tj}^2 / 2\Omega_j^2$. Eq. (7.7) has been generalised to various ionic species in the plasma as opposed to the one-component electron plasma originally considered by Bernstein. Hence, a study of the behaviour in a two-component plasma, in particular the ion-acoustic modes and ion Bernstein resonances, can be performed. In actual fact, the one-component version of the dispersion equation for plasma oscillations, which is obtained by equating the one-component version of the above longitudinal dielectric response function to zero, was first obtained by Gordeyev [18]. Nowadays, this form is seldom used since the exponential factor with the cosine function is replaced by an infinite series over the order of modified Bessel functions resulting in the summation over n in Eq. (7.3). However, we shall write this factor as

$$\exp\left(-\lambda_j\left[1 - \frac{1}{2}y^2 - \cos(y)\right]\right) = \sum_{k=0}^\infty (-1)^k y^{2k} d_k(\lambda_j) \;, \qquad (7.8)$$

where the polynomials, $d_0(x)=1$, $d_1(x)=0$, $d_2(x)=x/4!$, and $d_3(x)=x/6!$. The remaining polynomials can be determined by using either the novel graphical technique in the Appendix of Ref. [15] or the recurrence relation of

$$(2k+2)\, d_{k+1}(x) = x \sum_{j=0}^{k-1} \frac{d_j(x)}{(2k-2j+1)!} \;. \qquad (7.9)$$

The above equation has been obtained by differentiating Eq. (7.8). Then we see that the $d_k(x)$ are of order $[k/2]$, where $[x]$ represents the greatest integer less than or equal to x. They also obey the following equation:

$$\sum_{j=0}^k d_j(x)\, d_{k-j}(-x) = 0 \;. \qquad (7.10)$$

Since we can write the polynomials as $d_k(x) = \sum_{j=1}^{[k/2]} d_k^j \, x^j$ for $k \geq 2$, the recurrence relation reduces to

$$(2k+2)d^m_{k+1} = \sum_{j=0}^{k-1} \frac{d^{m-1}_j}{(2k-2j+1)!} \; . \tag{7.11}$$

On the other hand, general results for the coefficients of the polynomial can be obtained via the graphical method [13, 14], some of which are

$$d^1_k = \frac{1}{(2k)!} \; , \tag{7.12}$$

$$d^2_k = \frac{(2^{2k-2} - 1 - k(2k-1))}{(2k)!} \; , \tag{7.13}$$

$$d^{[k/2]}_k = \frac{1+(-1)^k}{2(k/2) \cdot (4!)^{k/2}} + \frac{1-(-1)^k}{2((k-3)/2)! \cdot 6! \cdot (4!)^{(k-3)/2}} \; . \tag{7.14}$$

As our aim is to consider the static limit, we can re-write Eq. (7.7) as

$$\epsilon_L(\mathbf{q},\omega) = 1 + \sum_j \frac{q^2_{Dj}}{q^2} - \sum_j \frac{q^2_{Dj}}{q^2} \int_0^\infty dy \left[\frac{d}{dy} \exp\left(-i\frac{\omega}{\Omega_j}y\right) \right]$$

$$\times \exp\left(-\frac{1}{2}\mu_j y^2 - \lambda_j[1-\cos(y)]\right) \; . \tag{7.15}$$

If one introduces Eq. (7.8) into the above equation and integrates by parts, then in the static limit one obtains

$$\epsilon_L(\mathbf{q},0) \equiv 1 - \sum_j \frac{q^2_{Dj}}{q^2} \sum_{k=0}^\infty (-1)^k d_k(\lambda_j) \int_0^\infty dy \frac{d}{dy} \left[y^{2k} \exp\left(-(\mu_j+\lambda_j)y^2\right) \right] \; . \tag{7.16}$$

This is a stunning result because all the terms other than $k=0$ vanish. As a consequence, Eq. (7.16) reduces to

$$\epsilon_L(\mathbf{q},0) = 1 + \sum_j \frac{q^2_{Dj}}{q^2} \; . \tag{7.17}$$

In other words, the static screening potential in a magnetised Maxwellian plasma is simply the isotropic Debye screening potential for any size of the magnetic field. This is really unlike the situation for the $T=0$ K CBG, where the static screening potential does acquire magnetic field corrections [13].

7.5. Ion-acoustic modes

So far, we have been concerned with the collective behaviour in the magnetised one-component Maxwellian plasma. Since the forms for the longi-

tudinal dielectric response function in weak and strong magnetic fields in
Refs. [13] and [15] are different from those in the standard texts [1, 19],
the same will apply in evaluating the collective modes in a two-component
plasma, a topic also studied in Ch. 4 of Ref. [1]. Here we shall be concerned
with the dispersion relation of the low frequency ion-acoustic mode in a
weak magnetic field. This mode is especially interesting because it involves
both small and large argument expansions of the complementary error function for the electrons and ions, which are denoted by the subscripts e and
i respectively.

According to Ichimaru, ion-acoustic modes occur in the frequency do-
main of $|q_z|v_{Ti} \ll |\omega| \ll |q_z|v_{Te}$. Unfortunately, because he is only able to
consider strong magnetic fields for Eq. (7.3), he is forced to consider the
frequency region of $|\omega| \ll \Omega_i$, which cannot reduce to the dispersion relation
for acoustic modes in the field-free limit given by Eq. (4.60a) in Ref. [1].
Furthermore, the analysis for the strong magnetic field limit is based on
Bernstein's approach [3], which has been shown (Appendix E of Ref. [13])
not to give the correct results for the Bernstein resonances. Here, we shall
study the ion-acoustic mode for the weak magnetic field case leaving the
strong magnetic field case to be published elsewhere.

In the previous section, the calculation of the screening potential of a
test-particle was simplified drastically because ω vanished. For the ion-
acoustic mode, however, it is no longer zero and now evaluation of the
exponential integral in Eq. (7.7) yields the more complicated result of

$$
\epsilon(\mathbf{q}, \omega) \equiv 1 + \sum_j \frac{q_{Dj}^2}{q^2} - i\sqrt{\pi} \sum_j \frac{\omega\, q_{Dj}^2}{q^3 v_{Tj}} \sum_{k=0}^{\infty} \Omega^{2k} d_k(\lambda_j)
$$
$$
\times \frac{\partial^{2k}}{\partial \omega^{2k}} \left[\exp\left(-\frac{\omega^2}{q^2 v_{Tj}^2}\right) \operatorname{erfc}\left(\frac{i\omega}{q v_{Tj}}\right) \right] . \quad (7.18)
$$

In obtaining Equivalence (7.18), we have used Eq. (2.3.15.4) in Ref. [20],
while erfc(z) is the complementary error function. The $\equiv$ symbol has been
introduced to indicate that if the series is divergent, then it must be regularised to yield the correct value for the longitudinal dielectric response
function. Otherwise, it can be replaced by an $=$ sign. A discussion of
regularisation of a divergent series is presented in Ref. [21] and is due to
be published in a book [22]. With the aid of Rodrigues' Formula, viz.
Eq. (8.959.1) in Ref. [23], and Eq. (7.1.19) in Ref. [24], Eq. (7.18) can be

written as

$$
\epsilon_L(\mathbf{q}, \omega) \equiv 1 + \frac{q_{De}^2}{q^2} + \frac{q_{Di}^2}{q^2} - 2i \frac{\omega q_{De}^2}{q^3 v_{Te}} \sum_{k=0}^{\infty} \left(\frac{\Omega_e}{q v_{Te}} \right)^{2k} d_k(\lambda_e)
$$

$$
\times \left[\frac{\sqrt{\pi}}{2} H_{2k} \left(\frac{\omega}{q v_{Te}} \right) \exp\left(-\frac{\omega^2}{q^2 v_{Te}^2} \right) \operatorname{erfc}\left(\frac{i\omega}{q v_{Te}} \right) \right.
$$

$$
\left. + \ \Theta(k-1) g_k \left(\frac{\omega}{q v_{Te}} \right) \right] - i\sqrt{\pi} \frac{\omega q_{Di}^2}{q^3 v_{Ti}}
$$

$$
\times \sum_{k=0}^{\infty} \Omega^{2k} d_k(\lambda_i) \frac{\partial^{2k}}{\partial \omega^{2k}} \exp\left(-\frac{\omega^2}{q^2 v_{Ti}^2} \right) \operatorname{erfc}\left(\frac{i\omega}{q v_{Ti}} \right),
$$

$$(7.19)$$

where $H_k(x)$ and $\Theta(x)$ represent respectively the Hermite polynomials and the Heaviside step-function, while the $g_k(x)$ are special polynomials obtained from

$$
g_k(x) = \sum_{j=0}^{2k-1} \binom{2k}{j} (-1)^{k-j} i^j H_j(x) H_{2k-1}(ix) . \tag{7.20}
$$

Some $g_k(x)$ obtained via Mathematica [25] are presented in Table 7.1. From these results we see that the polynomials can be written as

$$
g_k(x) = 2^k i x \sum_{j=0}^{k-1} g_k^j x^{2j} . \tag{7.21}
$$

If the general form for the coefficients of the Hermite polynomials, which can be obtained from Eq. (8.950.2) in Ref. [23], is introduced into Eq. (7.20), then we find after some calculation that

$$
g_k^1 = (-1)^k 2^k \left((1/2)_k - k! \right) , \tag{7.22}
$$

$$
g_k^2 = (-1)^k 2^{k+1} \left((k+1)!/3 - (k+1/3)(1/2)_k \right) , \tag{7.23}
$$

and

$$
g_k^{k-1} = 2^{k-1} , \tag{7.24}
$$

where $(x)_k$ is the Pochhammer notation for $\Gamma(k+x)/\Gamma(x)$.

To derive the dispersion relation for the ion-acoustic mode, we require that $|\omega/q v_{Te}| \ll 1$ and $|\omega/q v_{Ti}| \gg 1$. Hence, for the electron component we require small $|x|$ expansions for $\operatorname{erfc}(ix)$, the Hermite polynomials and the $g_k(x)$. Because large $|x|$ expansions are not forthcoming for the last two quantities, the ion component has been left in terms of $\exp(-x^2) \operatorname{erfc}(ix)$ for

Table 7.1.   The polynomials $g_k(x)$ given by Eq. (7.20).

k	$g_k(x)$
1	$2ix$
2	$4ix(-5 + 2x^2)$
3	$8ix(33 - 28x^2 + 4x^4)$
4	$16ix(-279 + 370x^2 - 108x^4 + 8x^6)$
5	$32ix(2895 - 5280x^2 + 2352x^4 - 352x^6 + 16x^8)$
6	$64ix(-35685 + 83370x^2 - 50232x^4 + 11376x^6 - 1040x^8 + 32x^{10})$

it is much simpler to introduce the asymptotic expansion for this quantity and take derivatives in order to obtain the large $|x|$ expansion for the ion component.

For $|\omega/qv_{Te}| \ll 1$, $g_k(\omega/qv_{Te})$ can be approximated by $g_k^1 \omega/qv_{Te}$, where g_k^1 is given by Eq. (23). However, to evaluate the sum involving the $g_k(x)$ in Eq. (7.19), we require approximations for $d_k(\lambda_e)$. For the extremely small magnetic field case, i.e. the limit as $B \to 0$, $\lambda_e \to \infty$. Then we can approximate the $d_k(\lambda_e)$ by $d_k^{[k/2]} \lambda_e^{[k/2]}$. After a little algebra we find that

$$
\begin{aligned}
G\left(\frac{\omega}{qv_{Te}}, \frac{\Omega_e}{qv_{Te}}, \frac{q_\perp}{q}\right) &\equiv \sum_{k=1}^{\infty} \left(\frac{\Omega_e}{qv_{Te}}\right)^{2k} d_k(\lambda_e)\, g_k\left(\frac{\omega}{qv_{Te}}\right) \\
&= \frac{i\omega}{qv_{Te}} \sum_{k=1}^{\infty} \left(\frac{\Omega_e^2\, q_\perp^2}{3q^4 v_{Te}^2}\right)^k \left[\frac{\Gamma(2k+1/2)}{\Gamma(1/2)\Gamma(k+1)} - \frac{\Gamma(2k+1)}{\Gamma(k+1)}\right. \\
&\qquad \left. - \frac{2\Omega_e^2}{15q^2 v_{Te}^2}\left(\frac{\Gamma(2k+3/2)}{\Gamma(k)\Gamma(1/2)} - \frac{\Gamma(2k+2)}{\Gamma(k)}\right) + \cdots\right] \\
&\qquad + O\left(\frac{\omega^2}{q^2 v_{Te}^2}\right). \quad (7.25)
\end{aligned}
$$

All the series in the above result are divergent. Hence, they are meaningless until they are regularised via Borel summation. On the other hand, since their limiting point is zero, we can truncate them after a few terms, thereby obtaining an accurate approximation. Therefore, we find that

$$
G\left(\frac{\omega}{qv_{Te}}, \frac{\Omega_e}{qv_{Te}}, \frac{q_\perp}{q}\right) \approx \frac{i\omega}{qv_{Te}}\left[-\frac{5}{12}\frac{\Omega_e^2\, q_\perp^2}{q^4 v_{Te}^2} + \frac{\Omega_e^4 q_\perp^2}{12q^6 v_{Te}^4}\left(\frac{11}{5} - \frac{279}{24}\frac{q_\perp^2}{q^2}\right) + \cdots\right].
$$

$$(7.26)$$

If we introduce the general form for the Hermite polynomials, then the sum, where they appear in Eq. (7.19), becomes

$$F\left(\frac{\omega}{qv_{Te}}, \frac{\Omega_e}{qv_{Te}}, \frac{q_\perp}{q}\right) = \sum_{k=0}^{\infty} \left(\frac{\Omega_e}{qv_{Te}}\right)^{2k} d_k(\lambda_e) H_{2k}\left(\frac{\omega}{qv_{Te}}\right)$$

$$= \sum_{k=0}^{\infty} (-1)^k \left(\frac{\Omega_e}{qv_{Te}}\right)^{2k} d_k(\lambda_e) \frac{\Gamma(2k+1)}{\Gamma(k+1)}$$

$$+ O\left(\frac{\omega^2}{q^2 v_{Te}^2}\right) . \quad (7.27)$$

For $\lambda_e \gg 1$, the first sum on the right hand side of Eq. (7.27) can be written as

$$\sum_{k=0}^{\infty} (-1)^k \left(\frac{\Omega_e}{qv_{Te}}\right)^{2k} d_k(\lambda_e) \frac{\Gamma(2k+1)}{\Gamma(k+1)}$$

$$= 1 + \frac{1}{4} \frac{\Omega_e^2 q_\perp^2}{q^4 v_{Te}^2} \frac{1}{12} \frac{\Omega_e^4 q_\perp^2}{q^6 v_{Te}^4} \left(-1 + \frac{35}{8} \frac{q_\perp^2}{q^2}\right) + O\left(\frac{\Omega_e^6}{q^6 v_{Te}^6}\right) . \quad (7.28)$$

To complete the evaluation of the entire term involving the Hermite polynomials in Eq. (7.19), we need the convergent power series expansion for the complementary error function with the exponential factor. This is

$$iu(z) = i \exp\left(-z^2\right) \mathrm{erfc}(iz) = \sum_{k=0}^{\infty} \frac{(-1)^k z^{2k+1}}{\Gamma(k+3/2)} + i \exp\left(-z^2\right) . \quad (7.29)$$

Technically, the exponential should be expanded as a power series, but as it only contributes to the damping, we retain it. Up to first order in ω/qv_{Te}, which is the highest order we are considering for the $g_k(x)$, we finally arrive at

$$i\sqrt{\pi} F\left(\frac{\omega}{qv_{Te}}, \frac{\Omega_e}{qv_{Te}}, \frac{q_\perp}{q}\right) u\left(\frac{\omega}{qv_{Te}}\right) =$$

$$\left[1 + \frac{1}{4} \frac{\Omega_e^2 q_\perp^2}{q^4 v_{Te}^2} + \frac{\Omega_e^4 q_\perp^2}{q^6 v_{Te}^4} \left(-1 + \frac{35 q_\perp^2}{8 q^2}\right) + \cdots\right]$$

$$\times \left[\frac{2\omega}{qv_{Te}} + i\sqrt{\pi} \exp\left(-\frac{\omega^2}{q^2} v_{Te}^2\right) + \cdots\right] . \quad (7.30)$$

To obtain the contribution from the ions, we require the complete asymptotic expansion for the function $u(z)$, viz.

$$i\sqrt{\pi} u(z) \equiv \sum_{k=0}^{\infty} \frac{\Gamma(k+1/2)}{\Gamma(1/2) z^{2k+1}} + 2i\sqrt{\pi} S \exp\left(-z^2\right) , \quad (7.31)$$

where the Stokes multiplier [26] is given by

$$
S = \begin{cases} 0 \ , & -\pi < \arg z < 0 \ , \\ \frac{1}{2} \ , & \arg z = 0 \ , \\ 1 \ , & 0 < \arg z < \pi \ . \end{cases} \tag{7.32}
$$

In fact, the result given by Equivalence (7.31) was the first ever example of the Stokes phenomenon [27]. We also require the derivative of this result, but according to Ref. [28], it is generally not permissible to differentiate an asymptotic expansion. This, however, only applies when the expansion is incomplete due to truncating higher order terms and neglecting subdominant terms, which is not the case with Equivalence (7.31). Therefore, for Equivalence (7.19) we have

$$
i\sqrt{\pi} \, \frac{\partial^{2k}}{\partial \omega^{2k}} \, u\left(\frac{\omega}{q\,v_{Ti}}\right) \equiv \frac{q\,v_{Ti}}{\omega^{2k+1}} \sum_{j=0}^{\infty} \frac{\Gamma(j+1/2)}{\Gamma(1/2)} \frac{\Gamma(2j+2k+1)}{\Gamma(2j+1)} \left(\frac{q\,v_{Ti}}{\omega}\right)^{2j}
$$
$$
+ \frac{2i\sqrt{\pi}}{(q\,v_{Ti})^{2k}} \, S \, \exp\left(-\frac{\omega^2}{q^2} v_{Ti}^2\right) H_{2k}\left(\frac{\omega}{q\,v_{Ti}}\right). \tag{7.33}
$$

If we introduce the above result into the ion component, or into final term of Equivalence (7.19), then we find that

$$
i\sqrt{\pi} \, \frac{\omega q_{Di}^2}{q^3 v_{Ti}} \sum_{k=0}^{\infty} \Omega_i^{2k} d_k(\lambda_i) \frac{\partial^{2k}}{\partial \omega^{2k}} \, u\left(\frac{\omega}{q v_{Ti}}\right) = \frac{q_{Di}^2}{q^2} \sum_{k=0}^{\infty} \left(\frac{\Omega_i}{\omega}\right)^{2k} d_k(\lambda_i)
$$
$$
\times \left[(2k)! + \frac{(2k+2)!}{4} \left(\frac{q v_{Ti}}{\omega}\right)^2 + \cdots \right.
$$
$$
\left. \cdots + 2i\sqrt{\pi} \left(\frac{\omega}{q v_{Ti}}\right)^{2k} \exp\left(-\omega^2/q^2 v_{Ti}^2\right) H_{2k}\left(\frac{\omega}{q v_{Ti}}\right) \right]. \tag{7.34}
$$

This result can be truncated for $\Omega_i \ll \omega$ which yields for the real part,

$$
\Re\left\{ i\sqrt{\pi} \, \frac{\omega q_{Di}^2}{q^3 v_{Ti}} \sum_{k=0}^{\infty} \Omega_i^{2k} d_k(\lambda_i) \frac{\partial^{2k}}{\partial \omega^{2k}} \, u\left(\frac{\omega}{q v_{Ti}}\right) \right\}
$$
$$
= \frac{q_{Di}^2}{q^2} \left[\sum_{j=0}^{\infty} \frac{\Gamma(j+1/2)}{\Gamma(1/2)} \left(\frac{q v_{Ti}}{\omega}\right)^{2j} \right.
$$
$$
+ \frac{q_\perp^2 v_{Ti}^2}{48} \frac{\Omega_i^2}{\omega^4} \sum_{j=0}^{\infty} \frac{\Gamma(j+1/2)}{\Gamma(1/2)} \frac{\Gamma(2j+5)}{\Gamma(2j+1)} \left(\frac{q v_{Ti}}{\omega}\right)^{2j} + \cdots \left.\right]
$$
$$
= \frac{q_{Di}^2}{q^2} \left[1 + \frac{q^2 v_{Ti}^2}{2\omega^2} + \frac{q^4 v_{Ti}^4}{\omega^4} \left(\frac{3}{4} + \frac{q_\perp^2}{q^2} \frac{\Omega_i^2}{q^2 v_{Ti}^2}\right) + O(\omega^{-6}) \right]. \tag{7.35}
$$

On the other hand, the imaginary part becomes

$$\Im\left\{ i\sqrt{\pi}\,\frac{\omega q_{Di}^2}{q^3 v_{Ti}}\sum_{k=0}^{\infty}\Omega_i^{2k} d_k(\lambda_i)\frac{\partial^{2k}}{\partial\omega^{2k}}\,u\left(\frac{\omega}{qv_{Ti}}\right)\right\} =$$
$$2i\sqrt{\pi}\,\frac{\omega q_{Di}^2}{q^3 v_{Ti}}\,S\,\exp\left(-\frac{\omega^2}{q^2}v_{Ti}^2\right)F\left(\frac{\omega}{qv_{Ti}},\frac{\Omega_i}{qv_{Ti}},\frac{q_\perp}{q}\right). \quad (7.36)$$

Previously, we were able to develop an expansion for $F(x,y,z)$ given by Eq. (7.28), but that was based upon $\omega \ll qv_{Te}$, which is not valid for the ion component. However, if we make the additional assumption that $\omega^2/q^2v_{Ti}^2 \ll qv_{Ti}/\Omega_i$, which is possible because the ion cyclotron frequency is significantly less than electron cyclotron frequency, then we can truncate the series in $F(x,y,z)$ after a few terms and introduce the forms for the Hermite polynomials and the $d_k(x)$. For example, neglecting the terms for $k \geq 3$ yields

$$F\left(\frac{\omega}{qv_{Ti}},\frac{\Omega_i}{qv_{Ti}},\frac{q_\perp}{q}\right) \approx 1 + \frac{q_\perp^2}{3q^2}\frac{\Omega_i^2\omega^4}{q^6 v_{Ti}^6} - \frac{q_\perp^2}{q^2}\frac{\Omega_i^2\omega^2}{q^4 v_{Ti}^4} + \frac{2q_\perp^2}{45q^2}\frac{\Omega_i^4\omega^6}{q^{10} v_{Ti}^{10}}. \quad (7.37)$$

This approximation is by no means ideal since there may be competing terms from larger values of the truncated value of k that have been neglected.

Now we are in a position to present the dispersion equation for the ion-acoustic mode with magnetic field corrections. By introducing Eqs. (7.26), (7.30), (7.35), and (7.36) into Equivalence (7.19), retaining the lowest order terms and setting them equal to zero, we obtain the dispersion equation of

$$1 + \frac{q_{De}^2}{q^2} - \frac{\omega^2 q_{De}^4}{\omega_{pe}^2 q^4}\left(1 + \frac{2}{3}\frac{\Omega_e^2 q_\perp^2}{q^4 v_{Te}^2}\right) - \frac{\omega_{pi}^2}{\omega^2} - \frac{\omega_{pi}^4 q^2}{\omega^4 q_{Di}^2}\left(3 + \frac{4q_\perp^2}{q^2}\frac{\Omega_i^2}{q^2 v_{Ti}^2}\right)$$
$$+ i\sqrt{\pi}\,\frac{\omega q_{De}^2}{q^3 v_{Te}}\left(1 + \frac{\Omega_e^2 q_\perp^2}{4q^4 v_{Te}^2}\right)\exp\left(-\frac{\omega^2}{q^2}v_{Te}^2\right)$$
$$+ 2i\sqrt{\pi}\,\frac{\omega q_{Di}^2}{q^3 v_{Ti}}\,S\exp\left(-i\frac{\omega^2}{q^2}v_{Ti}^2\right)\left(1 + \frac{q_\perp^2}{3q^2}\frac{\Omega_i^2\omega^4}{q^6 v_{Ti}^6}\right) \approx 0. \quad (7.38)$$

In the above result $-\omega$ has been replaced by ω, which ensures that the damping is given by a negative value as opposed to a positive value for Eq. (7.7). The dispersion relation for the ion-acoustic mode is obtained by solving the real part of the above result, which yields a cubic equation in ω^2. In principle, the cubic equation can be solved by following the method on p. 17 of Ref. [24], but it is very unwieldy. To simplify the calculation, however, we consider the cases of

(1) $\omega^6 \ll 3q^6 v_{Te}^2 v_{Ti}^2 T_e / 2m_i$,

(2) $\omega^6 \gg 3q^6 v_{Te}^2 v_{Ti}^2 T_e / 2m_i$.

For $\omega/q \sim \sqrt{T_e/m_i}$, the first case means that $T_e/T_i \ll 6m_i/m_e$, while the second case means that $T_e/T_i \gg 6m_i/m_e$. Though possible, the latter case is improbable.

In the first case we initially neglect the third term on the left hand side of the above approximation, while in the second case the fifth term is neglected initially. Thus, in both cases we solve a quadratic in ω^2 and perturb around these solutions to obtain the solution to the original cubic equation in ω^2. Unfortunately, when this method is applied to the second case, we find that the perturbation is of the same order as the leading term in the solution for the quadratic, which invalidates the second case. Hence, for this case we need to solve the original cubic dispersion equation. Therefore, only the first case yields the dispersion relation for the ion acoustic mode via the proposed perturbative method.

The dispersion relation for the first case is found to be

$$\omega^2 \approx \frac{1}{2} \frac{\omega_{pi}^2}{\alpha} \left(1 + \sqrt{1 + \frac{12\alpha\beta_1 q^2}{q_{Di}^2}} \right) + \frac{\omega_{pi}^4 q_{De}^4}{2\omega_{pe}^2 q^4 \alpha^3} \left(1 + \frac{2\Omega_e^2 q_\perp^2}{3q^4 v_{Te}^2} \right) \; , \quad (7.39)$$

where $\alpha = 1 + q_{De}^2/q^2$ and $\beta_1 = 1 + 4q_\perp^2 \Omega_i^2 / 3q^4 v_{Ti}^2$. For $q^2 \ll q_{De}^2$, the above result reduces to

$$\omega^2 \approx \left(\frac{T_e}{m_i} + \frac{3T_i}{m_i} + \frac{T_e m_e}{2m_i^2} \right) q^2 + \frac{13}{6} \frac{\Omega_i^2 q_\perp^2}{q^2} \; . \quad (7.40)$$

This is essentially the classic ion-acoustic mode with a correction due only to the ion cyclotron motion. In obtaining the above result it has been assumed that the electron and ion densities are equal to one another, while α has been replaced by q_{De}^2/q^2, which neglects terms that may compete with the magnetic correction. When T_e dominates, we see from Eq. (7.40) that $\omega \sim q\sqrt{T_e/m_i}$, which yields the condition given earlier for this case.

To obtain the damping, we put $\omega = \omega_r + i\gamma_i$ in the dispersion equation, Eq. (7.38), and then assume that $|\gamma_i/\omega_r| \ll 1$. Equating the imaginary parts to zero yields

$$\gamma_i \approx -i \frac{\sqrt{\pi}}{2} \frac{\omega_r^3}{\omega_{pi}^2 q} \left[\frac{q_{De}^2 \beta_3}{v_{Te}} + 2S \frac{q_{Di}^2 \beta_4}{v_{Ti}} \exp\left(-\frac{\omega_r^2}{q^2} v_{Ti}^2 \right) \right] \; ,$$

$$\text{where} \quad \beta_3 = 1 + \frac{\Omega_e^2 q_\perp^2}{4q^2 v_{Te}^2} \; , \quad \beta_4 = 1 + \frac{q_\perp^2 \Omega_i^2 \omega^4}{3q^8 v_{Ti}^6} \; , \quad (7.41)$$

and ω_r is given by Eq. (7.39).

In Ref. [15] it is shown that the Stokes multiplier S equals unity when $\gamma_i < 0$. This contradicts the damping result obtained by Landau for a Maxwellian plasma in his seminal paper [16], which corresponds to $S = 1/2$. If the electron and ion densities are equal to one another, then the above result for the damping reduces to Eq. (4-60a) in Ref. [1] for $B = 0$ and $S = 1/2$. However, since $S = 1$ due to the Stokes phenomenon, the equation should read

$$\omega \approx \omega_k \left\{ \pm 1 - i\sqrt{\frac{\pi}{8}} \left[\left(\frac{m_e}{m_i} \right)^{1/2} + 2 \left(\frac{T_e}{T_i} \right)^{3/2} \exp\left(-\frac{T_e}{2T_i} - \frac{3}{2} \right) \right] \right\}, \quad (7.42)$$

where $\omega_k = q\sqrt{T_e/m_i + 3T_i/m_i}$.

7.6. Conclusion

In this work further properties have been derived from the longitudinal dielectric response function for a magnetised Maxwellian plasma. First, the static screening potential for a test-particle in the system has finally been solved due to developments in asymptotics [13–15, 21, 22]. We have seen that regardless of the magnitude of the magnetic field, the static screening potential remains the Debye screening potential, i.e. there are no magnetic field corrections unlike the situation for the $T = 0$ K CBG in Ref. [13]. Next the dispersion relation for the ion-acoustic mode in a weakly magnetised two-component plasma was evaluated, which, until now, has only been presented for the field-free and strong magnetic field cases [1, 19]. In the future the results herein will be compared with the oscillations in a strongly magnetised two-component Maxwellian plasma using the appropriate asymptotic forms in Ref. [13].

References

[1] S. Ichimaru, *Basic Principles of Plasma Physics- A Statistical Approach*, W.A. Benjamin, Reading, MA, (1973).

[2] B. D. Fried and S. D. Conte, *The Plasma Dispersion Function*, Academic Press, New York, (1961).

[3] I. B. Bernstein, *Phys. Rev.* **109**, 10 (1958).

[4] A. E. Delsante and N. E. Frankel, *Ann. Phys. (N.Y.)* **125**, 240 (1980).

[5] D. F. Hines and N. E. Frankel, *Phys. Lett.* **A69**, 301 (1978).

[6] V. Kowalenko, N. E. Frankel, and K. C. Hines, *Phys. Rep.* **126(C)**, 109 (1985).

[7] N. S. Witte, R. L. Dawe, and K. C. Hines, *J. Math. Phys.* **22**, 1864 (1987).

[8] N. S. Witte, V. Kowalenko, and K. C. Hines, *Phys. Rev. D* **38**, 3667 (1988).

[9] S. R. Hore and N.E. Frankel, *Phys. Rev. B* **14**, 1952 (1976).

[10] V. Kowalenko and N. E. Frankel, *J. Math. Phys.* **35**, 6179 (1994).

[11] V. Kowalenko and N. E. Frankel, *Ann. Phys. (N.Y.)* **234**, 316 (1994).

[12] R. B. Dingle, *Asymptotic Expansions: Their Derivation and Interpretation*, Academic Press, London, (1973).

[13] V. Kowalenko, *Ann. Phys. (N.Y.)* **274**, 165 (1999).

[14] V. Kowalenko, *Acta Appl. Math.* **106**, 369 (2009).

[15] V. Kowalenko, *Phys. Lett.* **A337**, 405 (2005).

[16] L.D. Landau, *J. Phys.* **10**, 25 (1946).

[17] P.M. Morse and H. Feshbach, *Methods of Theoretical Physics Part I*, McGraw-Hill, New York, p. 609 (1953).

[18] G. V. Gordeyev, *JETP (USSR)* **6**, 660 (1952).

[19] D.G. Swanson, *Plasma Waves*, Academic Press, San Diego, CA (1989).

[20] A. P. Prudnikov, Yu. A. Brychkov and O. I. Marichev, *Integrals and Series Vol 1: Elementary Functions*, Gordon and Breach, New York (1986).

[21] V. Kowalenko, in *Recent Research Developments in Physics, Vol. 2*, (Ed. S.G. Pandalai), Transworld Research Network, Trivandrum, India, p. 17 (2001).

[22] V. Kowalenko, *The Stokes Phenomenon, Borel Summation and Mellin-Barnes Regularisation*, to be published by Bentham e-books.

[23] I. S. Gradshteyn and I. M. Ryzhik, *Table of Integrals, Series and Products, Fifth Ed.*, (Ed. A. Jeffrey), Academic Press, London (1994).

[24] *Handbook of Mathematical Functions*, (Eds. M. Abramowitz and I. A. Stegun), Dover, New York (1965).

[25] S. Wolfram, *Mathematica-A System for Doing Mathematics by Computer 2nd Ed.*, Addison-Wesley, Reading, MA (1991).

[26] M. V. Berry, *Proc. Roy. Soc. Lond.* **A 422**, 7 (1989).

[27] G. G. Stokes, *Collected Mathematical and Physical Papers Vol. 4*, Cambridge University Press, Cambridge, p. 77 (1904).

[28] E. T. Whittaker and G. N. Watson, *A Course of Modern Analysis*, Cambridge University Press, Cambridge, p. 153 (1973).

Chapter 8

The Boltzmann Equation in Fluorescent Lamp Theory

Graeme Lister

Osram Sylvania, CRSL,
71 Cherry Hill Dr., Beverly, MA 01915, USA
E-mail: Graeme.Lister@sylvania.com

Ken Hines was a wonderful mentor in my formative years as a physicist, and a great friend and companion throughout the rest of his life. He introduced me to the Fokker-Planck equation, which formed the basis of my Masters thesis, and showed me how to use it to derive the Boltzmann equation. Many years later, when my research took me into the field of gas discharge lighting, I was to re-discover the Boltzmann equation and apply it to fluorescent lamp modelling. Herein I discuss the important role the electron energy distribution function plays in understanding the physics of fluorescent lamps, and I describe some of the important insights gained from interpreting the Boltzmann equation.

8.1. Preamble

The day I first met Ken remains fixed in my memory as if it were yesterday. I had finished my summer job at Defence Standards Laboratories one Friday, and I appeared at his office door the next Monday, ready to start my postgraduate studies. Ken was a tall man, with a waistline that belied the enormous appetite I would later observe in our meals together. He had a clear, deep, resonant voice and his first question to me was "'Graeme, have you had a holiday?" When I replied in the negative, he said "Well, go away and have one, you're no damn use to me without having had a holiday!" So I spent a couple of weeks down on Victoria's Mornington Peninsula, before returning to Melbourne University suitably refreshed, and ready to begin my career in physics. I set to work understanding the Fokker-Planck equation, and from Ken, I learned to use it to derive the Boltzmann equation. After completing my Masters Thesis, I headed for the world of fully

ionised plasmas, and the "moments" of the Boltzmann equation at Flinders
University in South Australia, under the tutelage of Roger Hosking. After some twenty years in fusion research, in many different corners of the
globe, a career change took me into the world of discharge lighting, and I
was once again confronted with the Boltzmann equation in its purest form.
The article below represents part of the flowering tree grown from the seed
planted in the University of Melbourne, back in 1966. Perhaps not as elo-
quent as Ken's mathematical proof that 'tachyons cannot emit Cherenkov
radiation when passing through a bradyonic dielectric medium,' this is a
humble tribute to a man I admired greatly.

8.2. Introduction

In this paper, I discuss the role of the electron energy distribution function
(*EEDF*) in modelling of fluorescent lamps. Simple models of the *EEDF*,
together with experimental measurements of important parameters, can
add insight to the analysis of the power balance in low pressure discharge
lamps. In most cases of interest for fluorescent lamps, the assumption
that the *EEDF* is Maxwellian for electron energies lower the first excited
state is applicable. However, under some operating conditions, such as
low discharge current and low buffer gas pressure, this approximation is no
longer valid and a kinetic model should be used. Simple "two temperature"
models can also provide information about the depletion of "high energy"
tail of the *EEDF* by inelastic electron-atom collisions. As electron density
increases, the *EEDF* tends toward a Maxwellian distribution, even at high
electron energies, due to electron-electron collisions.

8.3. On low pressure discharge lamps

Gas discharges are used for a variety of lighting applications and cover a
wide range of plasma parameters. Examples can be found in each of the
major classifications of low temperature plasmas: "thermal" or *LTE* plasmas (High Intensity Discharge or *HID lamps* operating at gas pressures of
100–30,000 Pa, low pressure or *non-LTE* plasmas (e.g. fluorescent lamps
(*FL*), Low Pressure Sodium (*LPS*) lamps) operating at gas pressures typically between 40 and 1000 Pa, and "non-thermal" plasmas (Dielectric
Barrier Discharge (*DBD*) lamps), operating at pressures between 10 and
100 Pa. When an electric current is passed through a gaseous medium, it
produces a weakly ionised plasma. Electrical power is then converted into

heating of the gas, walls and electrodes of the discharge and radiation, some of which escapes from the discharge. Radiation is emitted by atoms and molecules of the gas which have been excited into electronic levels above the ground level, and then lose their energy by radiative decay to lower electronic levels. The fraction of electrical power converted to visible radiation (either directly or by use of a phosphor) determines the efficiency (or luminous efficacy, see Ref. [1] Sec. II) of the discharge as a light source. The underlying principle of low pressure discharges for light sources is the excitation of constituent atoms into radiation emitting states by electrons which are far from local thermal equilibrium (*LTE*) with the other species in the discharge. Of the three principal types of low pressure discharge lamps in use today, fluorescent lamps, low pressure sodium lamps and rare gas discharge lamps, fluorescent lamps are dominant in the marketplace. It has been estimated that 80% of the world's artificial light is fluorescent [2]. This paper describes some aspects of the theory and modelling of fluorescent lamps, which use mercury as the principal source of radiation. Fluorescent lamps are filled with a rare gas, typically argon at around 400 Pa pressure, with 0.5-5 Pa of mercury vapour. Under optimum conditions, 70-75% of electrical power in these discharges is converted to *UV* radiation by mercury atoms. The *UV* is then converted to visible light by means of a phosphor; the energy difference between the incoming *UV* photon and the outgoing visible photon (the Stokes shift) results in a total conversion efficiency of electrical power to visible light of about 25%. The electronic structure of the radiating atoms in the discharge is not the only criteria for efficient light production. The discharge must be able to produce a "white" light source, which illuminates objects in their "true" colours in comparison to a standard light source, such as sunlight. Ideally, the vapour pressure of the radiating atoms at close to room temperature should be sufficient to provide adequate light output. At room temperature, mercury has the highest vapour pressure of any of the elements suitable for producing radiation. The electronic levels of the mercury atom are shown in Fig. 8.1. Mercury is an efficient radiator because the energy level of the first excited state for resonance radiation is at 4.89 eV (254 nm), which is almost half of the ionisation energy (10.4 eV). This radiative level (and neighbouring metastable levels) are more readily populated by electron impact excitation than are the higher ones, providing efficient channels for both radiation and two step ionisation to maintain the discharge.

Table 1 lists some potential atomic candidates to be used as light sources, together with their excitation and ionisation energies, optimum

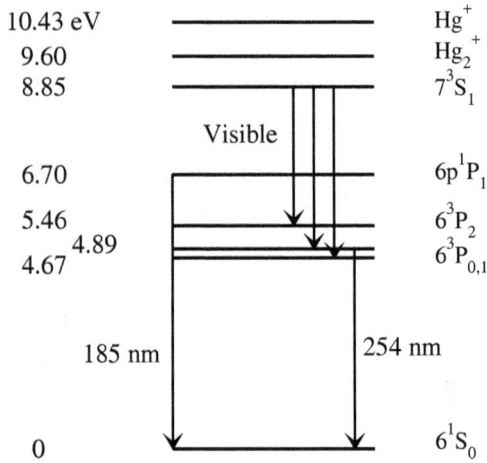

Fig. 8.1. Electronic levels of the mercury atom.

Table 8.1. Energy levels for excitation and ionisation, temperature of operation, and principal radiation wavelengths for a number of atoms investigated as potential light sources (*Ga introduced as GaI_3).

	ϵ_R (eV)	ϵ_i (eV)	T_c (°C)	λ_R (nm)
Ne	16.7	21.6	gas	74
Xe	8.45, 9.55	12.1	gas	147, 130
Hg	4.89, 6.71	10.4	40-55	254, 185
Na	2.10	5.14	260	589
Ba	2.24	5.22	400-600	554
Ba^+				455,493
				614,650
Ga	3.1	6.0	90-120*	407,413
GaI_3				390+

operating temperatures, and principal radiation wavelengths. Although Ne and Xe are gases at room temperature, the energy of the radiating states is well above half the ionisation energy. Further, radiation is emitted in the hard *UV*, which means a large fraction of energy is lost in converting the *UV* to visible radiation using a phosphor. As noted above, mercury has an ideal atomic structure, and the required vapour pressure is obtained at close to room temperature. The principal source of radiation is at 254 nm, and although this means about 50% of energy is lost in the conversion to visible light, the resulting efficiency of about 25% is superior to most other white light sources. The low pressure sodium lamp is the most efficient gas discharge light source known, but the yellow light emitted at 589 nm makes it unsuitable for general lighting, although it can be used for outdoor applications, where colour rendition is less important. Environmental issues regarding the use of mercury have motivated the lighting industry to seek other atomic and molecular radiators, for use in low pressure discharge lamps. Some years ago, Ba was extensively studied [3, 4]. As seen in Table 1, Ba has an ideal electronic structure, and the principal radiation is close to the peak of the eye sensitivity curve (555 nm), but the temperature required to maintain the discharge was far too high, and too much radiation was emitted in the infra-red. Unlike *HID* discharges, which operate at very high temperatures and are very short, low pressure discharges must be long, because the maintenance electric field is too small to provide sufficient electrical power in short discharges. Ga also has an ideal electronic structure, but Ga metal also has a low vapour pressure at near room temperatures. However, there have been encouraging results reported recently from discharges [5, 6], in which Ga is introduced in the form of GaI_3, which has a reasonable vapour pressure near 100°C, and dissociates in the discharge to form Ga atoms. Investigations into optimising fluorescent lamps have made extensive use of numerical modelling (See Ref. [1] Sec. VI.D and references therein), and models have also been used to understand discharges containing radiators other than mercury [4, 6]. These numerical models require an adequate representation of the *EEDF*, either using an analytic model or a numerical solution of the Boltzmann equation. The efficiency of conversion of electrical power to *UV* radiation depends on the fraction of high energy electrons in the *EEDF*, while the electrical characteristics are governed by the electrical conductivity, which is principally determined by the low energy electrons. Elastic collisions of electrons with atoms and ions couple the electric power to the discharge through the electrical conductivity and provide gas heating, while electron-electron collisions re-distribute the electron energy. In standard fluorescent lamps under normal operating conditions, the axial electron density is $n_c \sim 5 \times 10^{17} \mathrm{m}^{-3}$ and the ratio of electron density to gas density is $n_c/N \sim 5 \times 10^{-6}$. Under these conditions, electron-ion collisions play a minor role in establishing the electrical conductivity, but the situation is radically different at higher currents, particularly in "electrodeless" discharges, (see Sec. 5.2) in which rare gas pressures are typically an order of magnitude lower than in standard fluorescent lamps and electron densities are up to an order of magnitude higher. Under these conditions, the extremely high cross section for Coulomb collisions at low energies ensures that they play an important role (see Sec. 6). In low pressure gas discharge plasmas, the *EEDF* is almost never Maxwellian, at least for electron energies above the first excited level of the principal ionising species. This is because inelastic collisions of electrons with atoms cause the *EEDF* to be depleted in the inelastic energy range compared to that in the elastic energy range.

8.4. The physics of low pressure discharge lamps

The discharge in a conventional low pressure lamp is maintained between two electrodes. The main light producing region is the positive column, a region of constant electric field, separated from the cathode by the cathode dark space, the negative glow and the Faraday dark space, and from the anode by the anode dark space. All regions in the neighbourhood of electrodes contribute to inefficient use of power and this effect is minimised by ensuring that the positive column is much longer than the other regions of the discharge. For further details on the physics of low pressure discharge lamps see Ref. [1] Sec. VI, and references contained therein. The most important physical processes in the positive column are:

(1) *Electron-atom collision*: Elastic collisions of electrons with atoms and ions couple the electric power to the discharge through the electrical conductivity and provide gas heating, while electron-electron collisions re-distribute the electron energy and strongly influence the electron energy distribution function (*EEDF*). Collision processes between atoms in excited states and other atoms and molecules in the discharge can also play an important role, quenching radiative states and providing an extra channel for ionisation, which influences the current-voltage characteristics.

(2) *Radiation Transport*: Spectral lines emitted by atoms are broadened and shifted by a number of perturbing influences. The most important processes in low pressure discharge lamps are Doppler broadening, due to thermal motion of the radiating atom and collision resonance broadening resulting from perturbation of an atom radiating to the ground level by an identical atom in the ground level. Radiation emitted at one point in the discharge may be absorbed and re-emitted many times

before reaching the walls, broadening the radial density profiles of radiating states and consequently influencing the spectral output and power balance of the lamp. Radiation from the wings of each spectral line is less trapped than that from the center of the line, and for strongly absorbed lines the major contribution to the radiation emitted from the discharge is from the line wings. Radiation transport also influences other plasma properties, such as the maintenance electric field and the *EEDF*. For a detailed discussion of radiation transport in low pressure discharges, see Ref. [1] Sec. VI.B.2 and references therein.

(3) *Ambipolar Diffusion*: Charged particle diffusion plays an important role in low pressure discharge lamps by influencing the spatial distribution of atoms and molecules. Ion diffusion to the walls also represents an important loss mechanism in the power balance in the positive column. Electrons are much more mobile than ions and the ambipolar space charge field is established to maintain an equal radial flow of ions and electrons to the walls and preserve charge neutrality. Ions are thus accelerated away from the center of the discharge, while the electron motion is retarded.

The total electrical power W_{elec} in a gas discharge is dissipated through radiation (W_{rad}), heat conduction (W_{heat}), diffusion of particles from the discharge (W_{diff}) and acceleration of ions in the sheaths at the walls and electrodes (W_{sheath}), i.e.

$$W_{elec} = W_{rad} + W_{heat} + W_{diff} + W_{sheath}. \qquad (8.1)$$

Gas discharges for lighting are designed to maximise the fraction of electrical power emitted as visible radiation, or *UV* radiation that can be converted to visible radiation by a phosphor.

8.5. Approximations to the Boltzmann equation

8.5.1. *Numerical models of the positive column*

Theoretical and numerical models of the positive column in fluorescent lamps have been developed for over 40 years (see Ref. [1] Sec. VI.D) with varying degrees of complexity. The aim of positive column models is to predict the electrical characteristics, power balance and radiation output in a fluorescent lamp as a function of discharge diameter, rare gas type and pressure, discharge current and mercury vapour pressure. This is achieved by solving Eq. (8.1) as a set of modules, each incorporating the necessary sub-modules to compute each term separately. Models have provided valuable insight into the performance of fluorescent lamps, but have yet to realize their full potential in the design and optimisation of these lamps, hampered to a great extent by the lack of reliable atomic data. In order to calculate the transport and rate coefficients necessary to solve Eq. (8.1), a reliable representation of the *EEDF* is required. The most detailed model of the positive column of Hg-Ar developed to date [7], includes a solution of the non-local Boltzmann equation and a non-local model for radiation transport. This has proved valuable in deducing the power balance in a number of fluorescent lamp discharges, but it is highly computer intensive. Much insight has been gained into the role of the *EEDF* by adopting simper models [1, 8], which are outlined below.

8.5.2. *The Boltzmann equation*

The "local" form of the homogeneous Boltzmann equation at any point in the discharge, including elastic, inelastic and super-elastic electron-neutral and electron-electron interactions (but ignoring inter atomic processes) may be written in the form [8],

$$
\frac{d}{d\epsilon}\left[\left\{a(\epsilon)\frac{d}{d\epsilon}+b(\epsilon)\right\}f(\epsilon)\right]
$$
$$
=\sum_{j<k}\frac{n_j}{N}\left[\epsilon_{jk}\,q_{jk}(\epsilon)\,f(\epsilon)\ -\ (\epsilon+\epsilon_{jk})\,q_{jk}\,(\epsilon+\epsilon_{jk})\,f\,(\epsilon+\epsilon_{jk})\right]
$$
$$
+\sum_{j<k}\frac{n_j}{N}\left[\epsilon_{jk}\,q_{kj}(\epsilon)\,f(\epsilon)\ -\ (\epsilon-\epsilon_{jk})\,q_{kj}\,(\epsilon-\epsilon_{jk})\,f\,(\epsilon-\epsilon_{jk})\right],\ (8.2)
$$

where N and n_j are the gas density and the density of atomic levels 'j' respectively, and $f(\epsilon)$ is the electron energy probability function (*EEPF*), defined by the normalisation

$$
\int_0^\infty \sqrt{\epsilon}\,f(\epsilon,r)\,d\epsilon = 1, \tag{8.3}
$$

$$
a(\epsilon) = \frac{(E/N)^2\,\epsilon}{3q_t(\epsilon)}\ +\ 2\,\epsilon^2\,q_e\,A_1(\epsilon)\frac{n_e}{N} \tag{8.4}
$$

$$
b(\epsilon) = 2\sum_j \frac{m_e\epsilon^2 q_j(\epsilon)n_j}{M_s N}\ +\ 2\,\epsilon^2\,q_e\,A_2(\epsilon)\frac{n_e}{N} \tag{8.5}
$$

$$A_1(\epsilon) = \frac{2}{3}\left[\int_0^\epsilon x^{\frac{3}{2}} f(x)\, dx + \epsilon^{\frac{3}{2}} \int_\epsilon^\infty f(x)\, dx\right] \qquad (8.6)$$

$$A_2(\epsilon) = \int_0^\epsilon x^{\frac{1}{2}} f(x)\, dx \qquad (8.7)$$

and

$$q_t(\epsilon) = q_{em}(\epsilon) + q_{inel}(\epsilon) + \frac{n_e}{N}\frac{q_e(\epsilon)}{\gamma_E} \qquad (8.8)$$

is the total electron transport cross section [8], where $q_{em}(\epsilon)$ is the total electron momentum transfer cross section, $q_{inel}(\epsilon)$ is the total inelastic cross section, $\gamma_E = 0.582$ for a singly ionised gas [8], and $q_e(\epsilon)$ is the Coulomb cross section [9],

$$q_e(\epsilon) = \frac{\pi}{\epsilon^2}\left[\frac{e^2}{4\pi\epsilon_0}\right]^2 \ln(\Lambda); \qquad \ln(\Lambda) = \frac{1}{2}\ln\left[\frac{n_e e^6}{9(4\pi)^2 (\epsilon_0 k\epsilon_e)^3}\right]. \qquad (8.9)$$

Since the *EEDF* can deviate from a Maxwellian, ϵ_e has the meaning of an effective electron temperature. Note the limiting cases for Eqs. (8.6) and (8.7) for $\epsilon \gg \epsilon_e$ are

$$A_1(\epsilon) \sim \epsilon_e \ ; \qquad A_2(\epsilon) \sim 1. \qquad (8.10)$$

The first term on the right hand side of Eq. (8.2) represents energy losses due to collisions out of the volume $\epsilon \pm \Delta\epsilon$, and the second and third terms are energy gains into the volume due to inelastic and super-elastic collisions respectively. The second terms in Eqs. (8.4) and (8.5) represent electron-electron collisions while elastic electron-atom collisions are represented by the first term on the right hand side of Eq. (8.5). It is useful to divide the *EEDF* in low pressure discharges into two distinct regions in energy space

(1) "low energy" or "bulk" electrons, with energies less than ϵ_1,
    (the energy of the first excited state of the major ionising species)
    and
(2) "high energy" or "tail" electrons, with $\epsilon > \epsilon_1$.

8.5.3. *The "bulk" EEPF* ($\epsilon \le \epsilon_1$)

For many practical low pressure discharges, it is reasonable to assume that the principal inelastic and super-elastic electron-atom collisions contributing to the Boltzmann equation are those involving the ground state of the atoms in the discharge, because the excited state densities are considerably smaller. In this case, for $\epsilon \leq \epsilon_1$, the local form of the Boltzmann equation may be written (cf. Eq. (8.2))

$$\frac{d}{d\epsilon}\left[\left(a(\epsilon)\frac{\partial}{\partial\epsilon}+b(\epsilon)\right)f(\epsilon)\right]=0, \tag{8.11}$$

which has the general solution

$$f(\epsilon)=C_N\,\exp\left[-\int_0^{\epsilon}\frac{b(x)}{a(x)}dx\right]. \tag{8.12}$$

If Coulomb collisions are the dominant terms in the coefficients $a(\epsilon)$ and $b(\epsilon)$, i.e.

$$a_1=\frac{(E/N)^2\epsilon}{3q_t(\epsilon)}\quad\ll\quad a_2=2\epsilon^2\,q_e\,A_1(\epsilon)\frac{n_e}{N}, \tag{8.13}$$

$$b_1=2\sum_{js}\frac{m_e\epsilon^2 q_j^s(\epsilon)n_j^s}{M_s N}\quad\ll\quad b_2=2\epsilon^2\,q_e\,A_2(\epsilon)\frac{n_3}{N}, \tag{8.14}$$

then

$$\frac{b(\epsilon)}{a(\epsilon)}=\frac{A_2(\epsilon)}{A_1(\epsilon)}. \tag{8.15}$$

Using this approximation and the definitions of $A_1(\epsilon)$ and $A_2(\epsilon)$ (cf. Eqs. (8.6) and (8.7), the solution to Eq. (8.2) is found, by inspection, to be the Maxwell distribution,

$$f^M(\epsilon)=\frac{2}{\sqrt{\pi\epsilon_c^3}}\,\exp\left(-\frac{\epsilon}{\epsilon_c}\right). \tag{8.16}$$

For discharges in which Eq. (8.16) is valid, the approximation to a Maxwell distribution provides valuable insight into the discharge physics, particularly in relation to the coupling of electrical power through the electrical conductivity.

8.5.4. High energy electrons ($\epsilon > \epsilon_1$): the "Lagushenko" approximation

Again assuming that the principal inelastic and super-elastic electron-atom collisions contributing to the Boltzmann equation are those involving the ground state of the atoms in the discharge, noting that the ground state density, $n_0^s \sim N_s$, the total gas density of species 's', Eq. (8.2) may be written [8],

$$\frac{d}{d\epsilon} \left[\left(a(\epsilon)\frac{\partial}{\partial\epsilon} + b(\epsilon) \right) f(\epsilon) \right] - \sum_{sj} \frac{N_s}{N} q_{0j}^s(\epsilon) \, \epsilon \, f(\epsilon) =$$

$$\sum_{sj} \frac{n_j^s}{N} q_{j0}^s \left(\epsilon - \epsilon_0^2 \right) \left(\epsilon - \epsilon_0^2 \right) f \left(\epsilon - \epsilon_0^2 \right). \quad (8.17)$$

From the principle of detailed balance,

$$\epsilon \, q_{kj}(\epsilon) = \frac{g_j^s}{g_k^s} \left(\epsilon + \epsilon_{jk} \right) q_{jk} \left(\epsilon + \epsilon_{jk} \right), \quad (8.18)$$

where g_j^s is the statistical weight of electronic state 'j' in atoms of species 's', Eq. (8.17) may then be written,

$$\frac{d}{d\epsilon} \left[\left(a(\epsilon)\frac{\partial}{\partial\epsilon} + b(\epsilon) \right) f(\epsilon) \right] =$$

$$\sum_{sj} \frac{N_s}{N} q_{0j}^s(\epsilon) \, \epsilon \left[f(\epsilon) - \frac{n_j^s g_0^s}{N_s g_j^s} f \left(\epsilon - \epsilon_{0j}^s \right) \right], \quad (8.19)$$

where the right hand side of Eq. (8.19) represents the balance of inelastic and super-elastic collisions. In addition to the assumptions made at the beginning of this section, the form of the *EEDF* used by Lagushenko depends on the following further approximations;

(1) both $a(\epsilon)$ and $b(\epsilon)$ are weakly varying functions of ϵ, such that the WKB approximation may be used (i.e. the gradients of $a(\epsilon)$ and $b(\epsilon)$ are zero)

(2) the inequalities Eqs. (8.13) and (8.14) are assumed, and since $\epsilon_1 \gg \epsilon_e$ (cf. Eq. (8.10)), the electron-electron collision terms may be simplified such that

$$\frac{b(\epsilon)}{a(\epsilon)} \sim \frac{1}{\epsilon_e} \quad (8.20)$$

(3) the *EEPF* for electrons having super-elastic collisions, from the principle of detailed balance, satisfies

$$f \left(\epsilon - \epsilon_{0j} \right) = \exp \left(\frac{\epsilon_{0j}}{\epsilon_e} \right) f(\epsilon) \quad (8.21)$$

This is exact for a Maxwellian *EEPF*, but ignores the depletion of the high energy tail by inelastic collisions (i.e. the contribution of the super-elastic collisions to the *EEPF* is underestimated). Using these approximations, Eq. (8.19) has an analytic solution,

$$f(\epsilon) = C_N \, \exp \left(- \int_0^\epsilon G(x) \, dx \right) \tag{8.22}$$

$$G(\epsilon) = \frac{1}{2\epsilon_e} + \left[\frac{c(\epsilon)}{a(\epsilon)} + \frac{1}{4\epsilon_e^2} \right]^{\frac{1}{2}} \tag{8.23}$$

$$c(\epsilon) = \sum_{s,j} \frac{N_s}{N} \, q_{0j}^s(\epsilon) \, f(\epsilon) \, \epsilon \, \left[1 - \frac{n_j^s}{n_j^{sB}} \right] \tag{8.24}$$

where C_N is a normalisation constant, determined from Eq. (8.3) and n_j^{sB} is the particle density for atoms of species 's' in electronic state 'j' assuming a Boltzmann distribution (i.e. if the positive column were in *LTE*):

$$n_j^{sB} = \frac{n_0^s g_j^s}{g_0^s} \, \exp\left(-\epsilon_{jk}\right) \, \sim \, \frac{N_s g_j^s}{g_0^s} \, \exp\left(-\epsilon_{jk}\right). \tag{8.25}$$

The term $c(\epsilon)/a(\epsilon)$ in Eq. (8.23) represents the depletion of the high energy tail by collision processes. Deviations from Boltzmann equilibrium lead to an increase in $c(\epsilon)$ and a greater depletion of the high energy tail, while an increase in E/N or electron density n_e reduce this depletion, and in the limit of high electron density, the *EEPF* becomes approximately Maxwellian.

8.6. Validity of the "two temperature" approach

8.6.1. *Standard fluorescent lamps*

A comprehensive set of Langmuir probe experiments was performed on standard fluorescent lamp discharges in the early 1960s [10], measuring electric field E, electron temperature T_e and electron density $n_e(r)$ as a function of radius. These experiments, conducted in discharges with T12 lamps having inner diameter (*ID*) = 36 mm, formed the basis for validating a number of numerical models, many of them using a two temperature approximation to the *EEDF*. It is instructive, therefore, to examine the validity of the approximations in Eqs. (8.13) and (8.14) which are the necessary conditions that the *EEPF* be Maxwellian for electron energies below the first excited state (4.7 eV). Values of a_2/a_1 (cf. Eq. (8.13)) and of b_2/b_1 (cf. Eq. (8.14)) have been calculated using the experimental measurements for a number of values of mercury vapour pressure, Ar gas density and discharge current [11], and the results are given in Tables 2 and 3, for electron energies (ϵ) of 1, 3 and 4.7 eV. The total electron momentum transfer cross

Table 8.2. Values of $\frac{a_2}{a_1}$ from Eq. (8.13) for standard FL [10].

I(A)	p_{Ar}(Pa)	p_{Hg}(Pa)	1 eV	3 eV	4.7 eV
0.1	400.0	0.8	4.1	3.6	3.8
0.4	400.0	0.8	18.1	16.2	17.4
0.8	400.0	0.8	45	39	37
0.4	1.33	0.8	1.8	0.17	0.08
0.4	13.3	0.8	2.2	0.49	0.41
0.4	53.6	0.8	4.7	2.2	2.2
0.4	400.0	0.18	23.6	23.4	24.5
0.4	400.0	3.6	18	25	24

Table 8.3. Values of $\frac{b_2}{b_1}$ calculated from Eq. (8.14).

I(A)	p_{Ar}(Pa)	p_{Hg}(Pa)	1 eV	3 eV	4.7 eV
0.1	400.0	0.8	200	7.1	1.8
0.4	400.0	0.8	700	26	6.4
0.8	400.0	0.8	1300	49	12
0.4	53.6	0.8	2000	86	32
0.4	400.0	0.18	850	3.1	7.6
0.4	400.0	3.6	825	34	9.1

sections were calculated using the published values for mercury [12] and for argon [13].

For argon pressures above 0.1 Pa and currents above 200 mA, the approximation is valid, as is the condition $b_2/b_1 \gg 1$. At lower mercury vapour pressures, the condition $a_2/a_1 \gg 1$ is satisfied, but the condition $b_2/b_1 \gg 1$ appears questionable. However, for standard (400 mA) operation at low argon pressures (< 0.06 Pa) the condition $a_2/a_1 \gg 1$ is strongly violated and the *EEDF* more closely resembles a Druyvesteyn distribution. For these conditions, any agreement between experiment and models which assume a Maxwellian distribution for low energy electrons must be regarded as fortuitous. It is interesting to note in passing that fluorescent lamps are more efficient at lower argon pressures, but ion bombardment on the cathode increases and lamp life is consequently reduced.

8.6.2. *"Highly loaded" fluorescent lamps*

Recent developments in fluorescent lamps have led to increased power loading. For example, compact fluorescent lamps and FL for back-lighting have

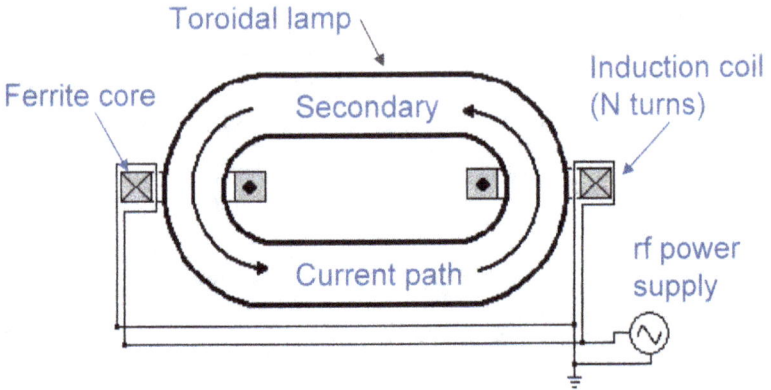

Fig. 8.2.   Schematic of the *ICETRON (ENDURA)*.

much narrower diameter than standard lamps, but operate with similar discharge currents: for example T2 lamps (*ID* = 5 mm) typically operate at 100 mA. Fluorescent lamps can also be operated at high power loading, using the principle of a ring discharge, similar to that of a tokamak in fusion research. In the *ICETRON*[a] lamp (*ENDURA*[a] in Europe) the discharge tube ring penetrates a ferrite core, with a primary winding to which *RF* power is applied. The discharge tube ring provides the single turn secondary (see Fig. 8.2). "Electrodeless" lamps, such as *ICETRON*, operate at argon pressures of a few tens of Pa, with discharge currents 7-12 A in a discharge equivalent to T16 (*ID* = 50 mm). In these cases, models have done less well in reproducing experimental results, in particular overestimating the maintenance electric field and hence the electrical power in the lamp [14].

Langmuir probe experiments have been performed on an *ICETRON* like lamp [15] and the experimental results used to compute values of a_2/a_1 and of b_2/b_1. The results are summarised in Tables 4 and 5. The essential parameters of the discharge are: *ID* = 50 mm, Hg pressure 0.8 Pa, Ar pressure 40 Pa. These tables clearly illustrate that the *EEDF* should be close to Maxwellian for electron energies below 4.7 eV and discharge currents above 2 A, in agreement with the experimental observations as shown in Fig. 8.3. Fig. 8.3 also shows that the high energy tail of the *EEDF* approaches a Maxwellian as current increases, due to the increase in electron-electron collisions. The *ICETRON* data were taken from Ref. [13].

---

[a]registered trademark

Table 8.4. Values of $\frac{a_2}{a_1}$, Eqs. (8.4 and (8.5), for *ICETRON*[15].

I(A)	p_{Ar}(Pa)	p_{Hg}(Pa)	1 eV	3 eV	4.7 eV
1	40	0.8	4.9	2.5	2.5
2	40	0.8	12	6	6
4	40	0.8	40	16	15
6	40	3.6	68	30	30

Table 8.5. Values of $\frac{b_2}{b_1}$, Eq. (8.4), for *ICETRON*[15].

I(A)	p_{Ar}(Pa)	p_{Hg}(Pa)	1 eV	3 eV	4.7 eV
1	40	0.8	3.1×10^3	121	28
2	40	0.8	6.0×10^3	234	57
4	40	0.8	1.1×10^4	523	139
6	40	3.6	2.0×10^4	760	190

8.7. Bulk electrons and electrical conductivity

An important part of modelling the behaviour of the positive column is a correct representation of the electric field as a function of the discharge parameters. This is represented by Ohm's law,

$$I = 2\pi E \int_0^R \sigma_e(r) \, r \, dr, \tag{8.26}$$

where I and E are the discharge current and electric field and σ_e is the electrical conductivity of the plasma,

$$\sigma_e = \frac{n_e(r)}{3N(r)} \sqrt{\frac{2e}{m_e}} \int_0^{} \frac{\epsilon}{q_t(\epsilon)} \frac{\partial f_0(r, \epsilon)}{\partial \epsilon} \, d\epsilon. \tag{8.27}$$

In Fig. 8.4, the experimentally measured discharge current is compared with that calculated from Eq. (8.26), using the Langmuir probe data for n_e and for two sets of experimental data, corresponding to the T12 [10] lamps and *ICETRON* [14]. Calculations used the experimental measurements of electron density and temperature for T12 from Verweij [10] and for *ICETRON* from Godyak *et al.* [15]. Results are compared with those obtained without Coulomb collisions (cf. Eq. (8.8)). These results show that even for the previously reported results for T12, there is considerable disagreement between the measured and calculated values. Earlier models ignored Coulomb collisions, so the previous good agreement between theory and experiment may have been fortuitous. The discrepancy is of such

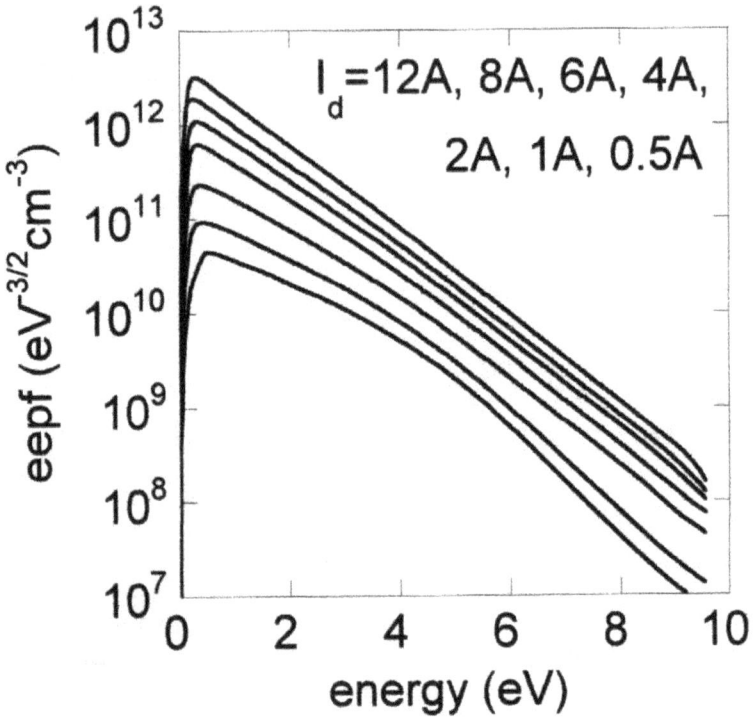

Fig. 8.3. Langmuir probe measurements of the *EEDF* as a function of discharge current in an *ICETRON* like discharge with $ID = 50$ mm, $p_{Ar} = 40$ Pa, and $p_{Hg} = 0.8$ Pa.

a fundamental nature that it brings into question the validity of applying Langmuir probes for these discharge parameters.

Calculations used the experimental measurements of electron density and temperature for T12 from Verweij [10] and for *ICETRON* from Godyak *et al.* [15].

8.8. Conclusions

The main purpose of this paper has been to discuss the role of the electron energy distribution function in modelling of fluorescent lamps. In most cases of interest for fluorescent lamps, the assumption that the *EEDF* is Maxwellian for electron energies lower the first excited state is applicable. However, under some operating conditions, such as low discharge current and low buffer gas pressure, this approximation is no longer valid and a

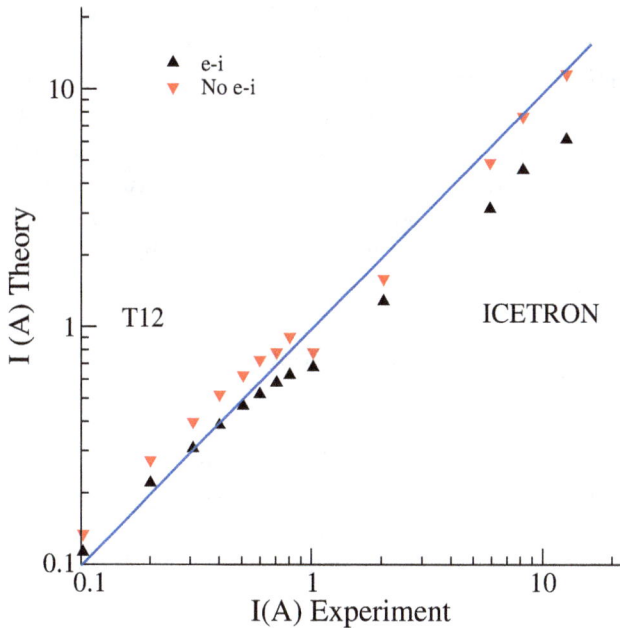

Fig. 8.4. Comparison of discharge currents in T12 and *ICETRON* calculated from Eq. (8.26) with the measured values.

kinetic model should be used. Since the validity of the model can be checked *a posteriori*, it is useful to include such a check in the computer program. Simple models of the *EEDF*, together with experimental measurements of important parameters, can add insight to the analysis of the power balance in low pressure discharge lamps [14]. Programs which solve the Boltzmann equation, Eq. (8.7) are more complex and are more computer intensive, but they can provide further details on the relative importance of fundamental processes, but they are often hampered by the lack of reliable atomic and molecular data.

Acknowledgments

I first encountered Ken in the pages of the Thursday edition of the Melbourne Age, where he penned a weekly science article. Ken had taken over from Keith Mather about the time I was taking an active interest in physics. Ken wrote with great clarity, and although I had not met him, his voice seemed to ring out from those pages and excite this teenage mind into thoughts of a career in physics. The article that convinced me described the then relatively new field of thermonuclear energy. I read of the exciting developments around the world in harnessing the energy used by the sun to generate heat, and use it for energy production on earth. This was in 1960, when "clean energy" and "global warming" were not on anyone's agenda, but Ken's words convinced me that the world's natural energy resources would eventually run out, and magnetic fusion seemed the way to provide an endless supply of clean energy. Many problems needed to be overcome, as is still the case today, but in my teenage fantasy, I would be the one to solve them!

Walking down Lygon St. near the University recently, I half expected to encounter a familiar figure, sauntering on his way to luncheon. Ken Hines bestrode the streets of Carlton like a colossus; this was his fiefdom, and he was known in every restaurant within 'coo-ee' of Melbourne University. Bon viveur, raconteur, lover of life, Ken was as happy with a good plate of gnocchi as with a tachyon, a glass of good wine as a Schwarzschild black hole, a Mozart sonata as a pion-nucleon pair. I was fortunate to share in but three of these passions.

References

[1] G. G. Lister, J. E. Lawler, W. Lapatovich, and V. Godyak, *Rev. Mod. Phys.* **76**, 541 (2004).

[2] M. G. Abeywickrama, in *Lamps and Lighting*, (eds. J. R. Coaton and A. M. Marsden), Arnold, London pp. 194 (1997).

[3] X. L. Peng, J. J. Curry, G. G. Lister, and J. E. Lawler, *J. Appl. Phys.* **91**, 1761 (2002).

[4] J. Laski, G. G. Lister, F. Palmer, P. E. Moskowitz, and J. J. Curry, *J. Appl. Phys.* **91**, 1772 (2002).

[5] D. J. Smith, J. D. Michael, V. Midha, G. M. Cotzas and T. J. Sommerer, *J. Phys. D: Appl. Phys.* **40**, 3842 (2007).

[6] S. Adamson *et al.*, *J. Phys. D: Appl. Phys.* **40**, 3857 (2007).

[7] G. M. Petrov and J. L. Giuliani, *J. Appl. Phys.* **94**, 62 (2002).

[8] R. Lagushenko and J. Maya, *Adv. in Atomic, Molecular and Optical Phys.* **26**, 321 (1990).

[9] L. Spitzer, in *Physics of Fully Ionized Gases* (second edition), Interscience, New York, (1961).

[10] W. Verweij, *Philips Res. Rep. Sup.* **2**, 1 (1961).

[11] G. G. Lister, V. A. Sheverev, and D. Uhrlandt, *J. Phys. D: Appl. Phys.* **35**, 2586 (2002).

[12] J. P. England and M. T. Elford, *Aust. J. Phys.*, **44**, 647 (1991).

[13] A. V. Phelps, unpublished data, ftp://jila.colorado.edu/collison.data.

[14] J. J. Curry, G. G. Lister, and J. E. Lawler, *J. Phys. D: Appl. Phys.* **35**, 2945 (2002).

[15] V. Godyak, R. Piejak, and B. Alexandrovich, in *Proceedings of the 9th International Symposium on the Science and Technology of Light Sources LS-9*, Cornell (ed. R. S. Bergman), Cornell University Press, Ithaca, NY, 157 (2001).

Chapter 9

Pair Modes in Relativistic Quantum Plasmas

D. B. Melrose* and J. McOrist[†]

School of Physics, University of Sydney
NSW 2006, Australia
*E-mail: melrose@physics.usyd.edu,au

Pair modes were identified by Hines and collaborators as longitudinal waves in a relativistic quantum plasma of spin 0 particles in the completely degenerate limit. Pair modes with both longitudinal and transverse polarization also exist in spin 1 plasmas in the completely degenerate limit. We consider the possible existence of pair modes in completely degenerate, spin $\frac{1}{2}$ plasmas, with a negative conclusion. It is argued that pair modes are associated with a pole (in the response function) that is present only in the presence of a Bose condensate. The properties of these wave modes, including their energetics, are discussed.

9.1. Introduction

When relativistic quantum effects are included in the electromagnetic response of a plasma[a], there is an additional source of dispersion. Besides the familiar dispersion associated with Landau damping (LD), there is also dispersion associated with one-photon pair creation (PC) and annihilation. Dissipation due to PC is possible only at high frequencies and for phase speeds greater than the speed of light, specifically above the threshold for a photon to decay into an electron-positron pair: $\omega > (4m^2 + |\mathbf{k}|^2)^{1/2}$ (we use natural units, $\hbar = c = 1$ in this paper). Dispersion associated with PC is nonzero at all frequencies, ω, and wave vectors, $\mathbf{k}$, but it has its largest effect in and near the range where PC is allowed. This additional dispersion leads not only to a modification of the properties of the natural wave modes

---

[†]Present address: Department of Physics, University of Chicago.
[a]DBM's interest in relativistic quantum plasmas developed independently of Ken Hines and his group, and the degree of overlap in published work is surprising, especially in view of the very small number of researchers in this field.

of the plasma that exist when PC is ignored, but also to the possibility of intrinsically new modes. One class of wave modes whose existence depends essentially on dispersion associated with PC are referred to as pair modes. The existence of pair modes in a degenerate (spin-0) boson gas was pointed out by Hines and colleagues [1].

In this paper we discuss the existence and properties of pair modes in completely degenerate gases, including the cases of spin 0, $\frac{1}{2}$ and 1. The only case identified by Hines and colleagues [1] was for a longitudinal mode in a spin 0 gas. Both longitudinal and pair modes have also been identified in a degenerate spin-1 gas [2]. The question arises as to whether pair modes exist in spin $\frac{1}{2}$ gas, which we refer to as an electron gas. We explore this possibility here. First we consider an artificial model in which the form of the response function is calculated for unpolarised spin $\frac{1}{2}$ particles and the occupation number is approximated by a δ-function, so implying that all particles have $\mathbf{p} = 0$. Pair modes exist in this model. However, the model is artificial because it is inconsistent with Fermi statistics. On searching for pair modes using the correct expression for the response function for a completely degenerate electron gas (with Fermi momentum $p_F \neq 0$) with relativistic effects included, we find no solutions. We argue that pair modes are associated with a pole in the response function, and that this pole is present only for Bose gases below their degeneracy temperature.

There is no simple physical model for a pair mode. As a step towards identifying a simple mode, we consider the energetics of pair modes. The wave properties for any wave mode include not only the dispersion relation and the polarization vector, but also the ratio of electric to total energy in the waves. This ratio suffices to determine the ratios of the mean energy in the electric and magnetic fields and in particle motions associated with the wave. The propagation of the wave energy is described by the group velocity, and the energy flux can be separated into the electromagnetic contribution (which is zero for longitudinal waves) and kinetic energy flux associated with the particle motions in the wave. The energetics suggest that a pair mode acts somewhat like a collection of dressed virtual pairs, but these cannot be real pairs because there is no Coulomb binding in the theory.

We discuss dispersion associated with PC in Sec. 9.2, and write down general expressions for the response tensors in Sec. 9.3. We identify pair modes in completely degenerate Bose gases and the apparent absence in degenerate Fermi gases is presented in Sec. 9.4. The wave properties of pair modes are discussed in Sec. 9.5 and conclusions are presented in Sec. 9.6.

9.2. Dispersion associated with PC

The classical resonance (Cerenkov) condition, corresponding to a wave resonating with a particle of velocity $\mathbf{v}$, is $\omega - \mathbf{k} \cdot \mathbf{v} = 0$. In a quantum treatment this condition arises from conservation of energy and momentum for a particle emitting or absorbing a wave quantum. The exact condition includes the quantum recoil, and the classical resonance condition corresponds to neglecting this, and retaining only the first order terms in an expansion in $\hbar$. The resonance condition can be satisfied only for $\omega/|\mathbf{k}| \leq |\mathbf{v}| < 1$, which defines the boundary of the allowed region for LD. The relativistic quantum generalisation of the resonance condition includes PC, which is allowed only for $\omega^2 > 4m^2 + |\mathbf{k}|^2$.

9.2.1. *Conservation of energy and momentum*

Consider a particle with initial 4-momentum p^μ (Greek indices run over 0–4, metric tensor with signature -2, $pk = p^\mu k_\mu = \varepsilon\omega - \mathbf{p} \cdot \mathbf{k}$) that either emits a wave quantum with 4-momentum k^μ, leaving it in the final state p'^μ, or absorbs a wave quantum with 4-momentum k^μ, leaving it in the final state p''^μ. Conservation of 4-momentum requires

$$p' - p + k = 0, \qquad p'' - p - k = 0, \qquad p''^2 = p'^2 = p^2 = m^2. \qquad (9.1)$$

On squaring to eliminate p' or p'', one obtains

$$ku - k^2/2m = 0, \qquad ku + k^2/2m = 0, \qquad (9.2)$$

where $u^\mu = p^\mu/m$ is the 4-velocity, with $ku = \gamma(\omega - \mathbf{k} \cdot \mathbf{v})$. The terms $\pm k^2/2m$ are quantum corrections (the 'recoil' terms), and when they are neglected, the classical resonance condition, $ku = 0$, is reproduced for either emission or absorption. The conditions, Eq. (9.2), also include an intrinsically relativistic quantum effect: the condition for one-photon pair creation and one photon pair annihilation. These actually correspond to crossed processes, described for example by $p'^\mu \rightarrow -p'^\mu$, so that $p'^\mu = k^\mu - p^\mu$ is the 4-momentum of the antiparticle (assuming p^μ describes the particle). All the resonance conditions may be incorporated into the single condition, cf. Eq. (9.2),

$$(ku)^2 - (k^2/2m)^2 = 0. \qquad (9.3)$$

In the next section, the response tensors for relativistic quantum plasmas are written in a form in which $(ku)^2 - (k^2/2m)^2$ appears as a resonance denominator.

9.2.2. *Threshold values for resonance*

One can determine the energy, momentum and velocity of a resonant particle by solving Eq. (9.3). There are four solutions in general, and two of these are related to the other two by $\omega \to -\omega$. It is convenient to write $\omega \to \epsilon\omega$, with $\epsilon = \pm 1$, so that only two solutions need to be written down explicitly. Of particular relevance are the values that correspond to the threshold condition $\mathbf{k} \cdot \mathbf{v} = \pm|\mathbf{k}|\,|\mathbf{v}|$. The values at this limit correspond to the threshold for either LD or PC.

It is convenient to introduce the parameters

$$a = \epsilon\frac{\left[\omega^2 - |\mathbf{k}|^2\right]}{2m\omega}, \qquad b = \epsilon\frac{|\mathbf{k}|}{\omega}. \tag{9.4}$$

For $|\mathbf{k} \cdot \mathbf{v}| = |\mathbf{k}|\,|\mathbf{v}|$ there are two solutions for each of the energy, $\varepsilon = \varepsilon_{\pm}$, momentum, $|\mathbf{p}| = p_{\pm}$, and speed, $|\mathbf{v}| = v_{\pm}$. These are given by

$$\frac{\varepsilon_{\pm}}{m} = \frac{a \pm b(a^2 + b^2 - 1)^{1/2}}{1 - b^2}, \qquad \frac{p_{\pm}}{m} = \frac{ab \pm (a^2 + b^2 - 1)^{1/2}}{1 - b^2},$$

$$v_{\pm} = \frac{p_{\pm}}{\varepsilon_{\pm}} = \frac{b \pm a(a^2 + b^2 - 1)^{1/2}}{a^2 + b^2}. \tag{9.5}$$

For a solution to be physically acceptable, $\varepsilon_{\pm} - m$, $p_{\pm}$, $v_{\pm}$ must all be real and positive. There may be zero, one or two physically acceptable solutions.

9.2.3. *Limits of the resonance regions*

The solutions of Eq. (9.5) for the energy may be written

$$\varepsilon_{\pm} = \frac{1}{2}\epsilon\omega \pm \varepsilon_k \qquad \varepsilon_k = \frac{1}{2}|\mathbf{k}|\left(\frac{\omega^2 - |\mathbf{k}|^2 - 4m^2}{\omega^2 - |\mathbf{k}|^2}\right)^{1/2}. \tag{9.6}$$

There are two allowed regimes determined by the conditions under which ε_k is real: $\omega^2 - |\mathbf{k}|^2 < 0$, which corresponds to LD, and $\omega^2 > 4m^2 + |\mathbf{k}|^2$, which corresponds to PC. We note two examples of the interpretation of the relevant two of the four solutions for a specific process at threshold. The first is the emission of a photon with $\varepsilon = \varepsilon_+$, $\epsilon = 1$ before emission and $\varepsilon' = \varepsilon_+$, $\epsilon = -1$ after emission. The second is pair creation with $\epsilon = 1$ and $\varepsilon_{\pm}$ being the energies of the particle and antiparticle. Pair modes depend on dispersion due to PC and only occur near the threshold $\omega^2 = 4m^2 + |\mathbf{k}|^2$.

9.3. Response tensor for a relativistic quantum gas

In this section we write down general forms for the response tensors for an electron gas, and for boson gases with spin 0 and spin 1. In all cases, the response tensor is written such that all resonances are included in a resonant denominator of the form $(ku)^2 - (k^2/2m)^2$, cf. Eq. (9.3). Also, in all cases particles and anti-particles contribute in the same way, and $\tilde{n}(\mathbf{p})$ denotes the occupation number summed over particle and anti-particle contributions.

9.3.1. *General properties of the response tensor*

The response tensor, $\mathcal{P}^{\mu\nu}(k)$, is defined here by analogy with the vacuum polarization. Specifically, $\mathcal{P}^{\mu\nu}(k)$ relates the induced 4-current density, $J^\mu(k)$, to the 4-potential, $A^\mu(k)$,

$$J^\mu(k) = \mathcal{P}^\mu{}_\nu(k)\,A^\nu(k), \qquad k_\mu \mathcal{P}^{\mu\nu}(k) = 0 = k_\nu \mathcal{P}^{\mu\nu}(k), \qquad (9.7)$$

where the constraints impose charge continuity and gauge invariance, respectively. The response tensor includes the vacuum polarization tensor. We have examined the effect of the vacuum polarization on the properties of pair modes, and found it to be insignificant; the vacuum polarization tensor is ignored here.

The response tensor involves an integral over the occupation number of the particles, and can be written in the form

$$\mathcal{P}^{\mu\nu}(k) = -e^2 \int \frac{\tilde{n}(\mathbf{p})}{\varepsilon} \frac{(ku)^2}{(ku)^2 - (k^2/2m)^2} N^{\mu\nu}(k,u) \frac{d^3\mathbf{p}}{(2\pi)^3}, \qquad (9.8)$$

where $N^{\mu\nu}(k,u)$ is different for electrons and bosons of spin 0 and spin 1.

The general form, Eq. (9.8) is constrained by the charge-continuity and gauge-invariance conditions, which restricts possible tensorial forms for $N^{\mu\nu}(k,u)$. There are only two independent second-rank tensors that can both be constructed from the metric tensor, $g^{\mu\nu}$, and the 4-vectors k^μ, u^μ, and also satisfy the charge-continuity and gauge-invariance conditions. One of these is $g^{\mu\nu} - k^\mu k^\nu/k^2$, which is independent of u^μ, and which is the only tensorial form allowed for the vacuum contribution. The other tensor is written here as

$$a^{\mu\nu}(k,u) = g^{\mu\nu} - \frac{k^\mu u^\nu + k^\nu u^\mu}{ku} + \frac{k^2 u^\mu u^\nu}{(ku)^2}. \qquad (9.9)$$

It follows that the numerator in the expression, Eq. (9.8), for the response tensor must be of the form

$$N^{\mu\nu}(k,u) = b(k,u)\,a^{\mu\nu}(k,u) + c(k,u)\left(g^{\mu\nu} - \frac{k^\mu k^\nu}{k^2}\right), \qquad (9.10)$$

where $b(k,u)$, $c(k,u)$ are different for electrons and bosons of spin 0 and spin 1.

It is possible to rearrange the terms in the numerator proportional to $g^{\mu\nu} - k^\mu k^\nu/k^2$ to include factors $(ku)^2 - (k^2/2m)^2$ that cancel with the resonant denominator. In particular, such a cancellation appears naturally in the transverse response functions for boson gases. After cancelling the resonant denominator, such terms are proportional to the proper number density, $\tilde{n}_{\rm pr}$, which is not to be confused with the number density, $\tilde{n}$, in the rest frame of the gas. For unpolarised particles (the only case considered here) of spin S, these number densities are given by

$$\tilde{n}_{\rm pr} = (2S+1)\int \frac{m}{\varepsilon}\,\tilde{n}(\mathbf{p})\frac{d^3\mathbf{p}}{(2\pi)^3}\,, \qquad \tilde{n} = (2S+1)\int \tilde{n}(\mathbf{p})\frac{d^3\mathbf{p}}{(2\pi)^3}\,. \qquad (9.11)$$

where the tilde denotes the sum over both particles and anti-particles.

9.3.2. *Isotropic plasmas*

For an isotropic medium, the response tensor may be separated into longitudinal and transverse parts. In this case the relevant 4-velocity, $\tilde{u}^\mu$, is that of the rest frame of the gas (which is the only frame in which it is isotropic). The separation corresponds to

$$\mathcal{P}^{\mu\nu}(k) = \mathcal{P}^L(k)\,L^{\mu\nu}(k,\tilde{u}) + \mathcal{P}^T(k)\,T^{\mu\nu}(k,\tilde{u}), \qquad (9.12)$$

with the the longitudinal and transverse projection tensors, $L^{\mu\nu}(k,\tilde{u})$, $T^{\mu\nu}(k,\tilde{u})$, related to the tensors introduced above by

$$g^{\mu\nu} - \frac{k^\mu k^\nu}{k^2} = \frac{(k\tilde{u})^2}{k^2}L^{\mu\nu}(k,\tilde{u}) + T^{\mu\nu}(k,\tilde{u}),$$
$$a^{\mu\nu}(k,\tilde{u}) = L^{\mu\nu}(k,\tilde{u}) + T^{\mu\nu}(k,\tilde{u}). \qquad (9.13)$$

9.3.3. *Response tensors for electrons and bosons*

Detailed calculations of the response tensors have been presented elsewhere [1–3], and here we simply state the results. In the general form, Eq. (9.8), when the response tensors are as defined in Eq. (9.10), the two invariants, $b(k,u)$, $c(k,u)$, have the values listed in Table 9.1 for an electron gas and for boson gases of spin 0 and spin 1.

Table 9.1. We tabulate, for spin 0, $\frac{1}{2}$ and 1, the invariants in the numerator of Eq. (9.10) in the response function in the form Eq. (9.8), and the invariants for the longitudinal and transverse response functions in the completely degenerate limit, as defined by Eq. (9.14).

Spin	$b(k,u)$	$c(k,u)$	$b^L(k)$	$c^T(k)$	$b^T(k)$
0	1	$-k^4/4m^2(ku)^2$	$1 - k^2/4m^2$	1	0
$\frac{1}{2}$	1	0	1	0	1
1	$1 - k^2/6m^2$	$-k^4/12m^2(ku)^2$	$1 - k^2/4m^2$	1	$2(1 - k^2/4m^2)$

In the completely degenerate case there are no anti-particles (in either a Fermi or a Bose gas). In a Bose gas the occupation number is replaced according to $(2S+1)\tilde{n}(\mathbf{p}) \to n(2\pi)^3\delta^3(\mathbf{p})$, corresponding to all the particles in their ground state. In this case one has $n_{\mathrm{pr}} = n$, and in an arbitrary frame all the particles have $p^\mu = m\tilde{u}^\mu$, where $\tilde{u}$ is the 4-velocity of the rest frame. The integral in Eq. (9.8) is then trivial, and the 4-velocity, u, of the particles is replaced by $\tilde{u}$ in the argument of $a^{\mu\nu}(k, u)$.

On setting $u = \tilde{u}$, the response tensor, Eq. (9.10), for a completely degenerate Bose gas reduces to the form in Eq. (9.12). The longitudinal and transverse response functions are of the form

$$\mathcal{P}^{L,T}(k) = -\frac{e^2 n}{m} \left[c^{L,T}(k) + \frac{(k\tilde{u})^2}{(k\tilde{u})^2 - (k^2/2m)^2} \, b^{L,T}(k) \right], \qquad (9.14)$$

with the $c^L(k) = 0$ for all cases considered here, and with $c^T(k)$, $b^{L,T}(k)$ given by the values in Table 9.1.

The response tensor, Eq. (9.14), contains a pole which is associated with the Bose condensate. At a nonzero temperature below the degeneracy temperature, a finite fraction of all particles are in the ground state, and the response tensor includes a contribution of the form in Eq. (9.14). There is also a contribution from the particles in excited states, and this does not contain a pole. This point is important in our argument that pair modes are intrinsic to degenerate Bose gases.

The completely degenerate model $\tilde{n}(\mathbf{p}) \propto \delta^3(\mathbf{p})$ is valid for bosons, but not for fermions, which have $\tilde{n}(\mathbf{p}) = 1$ for $|\mathbf{p}| < p_F$, and $\tilde{n}(\mathbf{p}) = 0$ for $|\mathbf{p}| > p_F$, where p_F is the Fermi momentum. In the limit $p_F \to 0$, the number density of the fermions goes to zero, $n \propto p_F^3$. The assumption $p_F = 0$ is artificial for fermions, and although we use it as an illustrative example that leads to pair modes, we find that the exact result in the limit $p_F \neq 0$ implies that no pair modes exist. This leads us to conclude that pair modes exist only for cases where the assumption $\tilde{n}(\mathbf{p}) \propto \delta^3(\mathbf{p})$ is well justified, specifically, plasmas that include Bose condensates.

9.3.4. *Jancovici's (1962) longitudinal response function*

The exact result for a completely degenerate relativistic electron gas was derived by Jancovici [1, 3–5]. The longitudinal response function is

$$\mathcal{P}^L(k) = \frac{e^2\omega^2}{4\pi^2|\mathbf{k}|^2} \left\{ \frac{8\varepsilon_{\mathrm{F}} p_{\mathrm{F}}}{3} - \frac{2|\mathbf{k}|^2}{3}\ln\left(\frac{\varepsilon_{\mathrm{F}} + p_{\mathrm{F}}}{m}\right) + T_1 + T_2 + T_3 \right\} \quad (9.15)$$

where

$$T_1 = \frac{\varepsilon_{\mathrm{F}}[4\varepsilon_{\mathrm{F}}^2 + 3(\omega^2 - |\mathbf{k}|^2)]}{6|\mathbf{k}|} \ln \left| \frac{4\varepsilon_{\mathrm{F}}^2\omega^2 - \left(\omega^2 - |\mathbf{k}|^2 - 2p_{\mathrm{F}}\,|\mathbf{k}|\right)^2}{4\varepsilon_{\mathrm{F}}^2\omega^2 - \left(\omega^2 - |\mathbf{k}|^2 + 2p_{\mathrm{F}}\,|\mathbf{k}|\right)^2} \right|,$$

$$T_2 = \frac{\omega\left[3|\mathbf{k}|^2 - \omega^2 - 12\varepsilon_{\mathrm{F}}^2\right]}{12|\mathbf{k}|} \ln \left| \frac{4(\varepsilon_{\mathrm{F}}\omega + p_{\mathrm{F}}\,|\mathbf{k}|)^2 - (\omega^2 - |\mathbf{k}|^2)^2}{4(\varepsilon_{\mathrm{F}}\omega - p_{\mathrm{F}}\,|\mathbf{k}|)^2 - (\omega^2 - |\mathbf{k}|^2)^2} \right|,$$

and

$$\begin{aligned}
T_3 = {} & \frac{2m^2 + \omega^2 - |\mathbf{k}|^2}{3(\omega^2 - |\mathbf{k}|^2)}|\mathbf{k}|\varepsilon_k \\
& \times \ln \left| \frac{(\omega^2 - |\mathbf{k}|^2)^2 (\varepsilon_{\mathrm{F}}|\mathbf{k}| + 2p_{\mathrm{F}}\varepsilon_k)^2 - 4m^4\omega^2|\mathbf{k}|^2}{(\omega^2 - |\mathbf{k}|^2)^2 (\varepsilon_{\mathrm{F}}|\mathbf{k}| - 2p_{\mathrm{F}}\varepsilon_k)^2 - 4m^4\omega^2|\mathbf{k}|^2} \right|. \quad (9.16)
\end{aligned}$$

Here, $\varepsilon_{\mathrm{F}} = (m^2 + p_{\mathrm{F}}^2)^{1/2}$. This function does not have poles at the thresholds for LD and PC. At these thresholds, ε_k becomes imaginary, and the final logarithmic factor is replaced by an arctangent [1, 5].

The transverse response function written down by Jancovici [4] involves the same logarithmic functions as Eq. (9.16), and the comments made here about pair modes in an electron gas apply to both longitudinal and transverse modes. However, Jancovici's expression for the transverse response function also contains a spurious overall factor that incorrectly implies a pole at $\omega^2 = |\mathbf{k}|^2$.

9.4. Pair modes in completely degenerate gases

Longitudinal and transverse wave modes correspond to solutions of the longitudinal and transverse dispersion equations. The wave modes are discussed here for completely degenerate, relativistic gases of spin 0, $\frac{1}{2}$ and 1.

9.4.1. *Dispersion relations*

The dispersion relations for longitudinal and transverse waves are

$$(k\tilde{u})^2 + \mu_0 \mathcal{P}^L(k) = 0, \qquad k^2 + \mu_0 \mathcal{P}^T(k) = 0, \tag{9.17}$$

respectively. The vacuum polarization contributes to both dispersion equations. The form of these contributions follows from the fact that the vacuum polarization tensor is proportional to $g^{\mu\nu} - k^\mu k^\nu / k^2$, which can be separated into longitudinal and transverse parts using Eq. (9.13). The corrections to Eq. (9.17) correspond to multiplying the terms $(k\tilde{u})^2$ and k^2 by a factor that differs from unity by a small amount proportional to the vacuum polarization tensor. This correction is unimportant here and we neglect it in the following discussion.

9.4.2. *Dispersion in relativistic quantum plasmas*

Before considering particular examples of dispersion in relativistic quantum plasmas, we note that the new features that appear, compared with the non-quantum, non-relativistic limit, are associated with the zeros of the resonant denominator, i.e. of Eq. (9.3). In the rest frame, $\tilde{u} = [1, \mathbf{0}]$, setting the denominator to zero gives $\omega^2 - (\omega^2 - |\mathbf{k}|^2)^2/4m^2 = 0$. Solving for ω^2 gives $\omega^2 = \omega_\pm^2$, with

$$\omega_\pm^2 = 2m^2 + |\mathbf{k}|^2 \pm 2m \left(m^2 + |\mathbf{k}|^2 \right)^{1/2} \approx \begin{cases} 4m^2 + 2|\mathbf{k}|^2, \\ |\mathbf{k}|^4/4m^2, \end{cases} \tag{9.18}$$

where the approximation is for $|\mathbf{k}|^2 \ll m^2$. The upper solution is just above the threshold for one-photon pair creation, and is characteristic of a pair mode. The lower solution is roton-like [6]. In all the case we consider, the dispersion relations for pair modes approach the solution $\omega^2 = \omega_+^2$ in the low-density limit, $\omega_p^2 \to 0$. This suggest that pair modes are directly associated with a pole at $\omega^2 - (\omega^2 - |\mathbf{k}|^2)^2/4m^2 = 0$, and only exist in plasmas for which the response function contains such a pole.

9.4.3. *Completely degenerate spin 0 gas*

The dispersion equation for longitudinal waves in a completely degenerate spin 0 gas is

$$\omega^4 - \omega^2 \left(2|\mathbf{k}|^2 + \omega_p^2 + 4m^2 \right) + |\mathbf{k}|^4 + |\mathbf{k}|^2 \omega_p^2 + 4m^2 \omega_p^2 = 0. \tag{9.19}$$

The two solutions for ω^2 are

$$\omega^2 = \frac{1}{2} \left(4m^2 + \omega_p^2 + 2|\mathbf{k}|^2\right) \pm \frac{1}{2} \left[(4m^2 - \omega_p^2)^2 + 16m^2|\mathbf{k}|^2\right]^{1/2}, \quad (9.20)$$

with $\omega_p^2 = e^2 n/\varepsilon_0 m$. These solutions simplify for $\omega_p^2 < 4m^2$, $|\mathbf{k}|^2 \ll 4m^2$ to

$$\omega^2 = 4m^2 + |\mathbf{k}|^2 + 4m^2|\mathbf{k}|^2/(4m^2 - \omega_p^2) + \cdots,$$
$$\omega^2 = \omega_p^2 + |\mathbf{k}|^2 - 4m^2|\mathbf{k}|^2/(4m^2 - \omega_p^2) + \cdots. \quad (9.21)$$

The former of these is the pair mode identified by Kowalenko, Frankel and Hines [1], and it reduces to Eq. (9.18) as expected for $\omega_p^2 \to 0$. The latter solution is a conventional Langmuir-like mode in the degenerate gas. In the limit $\omega_p^2 \to 0$ one needs to keep the term of order $|\mathbf{k}|^4$, and then the lower solution of Eq. (9.21) reduces to the corresponding solution in Eq. (9.18). The dispersion equation for transverse waves in the completely degenerate spin 0 gas has the familiar form $\omega^2 = \omega_p^2 + |\mathbf{k}|^2$. There is no pair mode with transverse polarization in a spin 0 plasma.

9.4.4. *Idealised cold quantum electron gas*

The simple model of a cold quantum electron gas, obtained by assuming a δ-function occupation number, leads to the dispersion equations

$$\left(\omega^2 - |\mathbf{k}|^2\right)^2 - 4m^2 \left(\omega^2 - \omega_p^2\right) = 0$$
$$\left(\omega^2 - |\mathbf{k}|^2\right)^3 - 4m^2\omega^2 \left(\omega^2 - \omega_p^2 - |\mathbf{k}|^2\right) = 0, \quad (9.22)$$

for longitudinal and transverse waves, respectively. Writing the two solutions of Eq. (9.22) for longitudinal waves as $\omega^2 = \omega_\pm^2(\mathbf{k})$, one finds

$$\omega_\pm^2(\mathbf{k}) = 2m^2 + |\mathbf{k}|^2 \pm 2m \left(m^2 - \omega_p^2 + |\mathbf{k}|^2\right)^{1/2}$$
$$\approx \begin{cases} 4m^2 + 2|\mathbf{k}|^2 - \omega_p^2, \\ \omega_p^2 + \frac{1}{4m^2} \left(\omega_p^2 - |\mathbf{k}|^2\right)^2, \end{cases} \quad (9.23)$$

where the approximate forms apply for $\omega_p^2, |\mathbf{k}|^2 \ll m^2$. The solution $\omega = \omega_-(\mathbf{k})$ reduces to $\omega = \omega_p$ when dispersion due to PC is unimportant, and this corresponds to the cold plasma approximation for Langmuir waves. The other solution has a cutoff at $\omega^2 = 2m \left[m + (m^2 - \omega_p^2)^{1/2}\right]$ just below the PC threshold, and would appear to be a pair mode. There are three solutions of the transverse dispersion relation. For $\omega_p^2, |\mathbf{k}|^2 \ll m^2$, one of these reduces to the familiar transverse waves in a cold plasma, one exists only near $\omega = 2m$ and is a pair mode, and the third exists only near $\omega = 0$ and is roton-like.

9.4.5. *No pair modes in degenerate electron gas?*

As already noted, the idealised model for a cold quantum electron gas obtained by assuming $n(\mathbf{p}) \propto \delta^3(\mathbf{p})$ is artificial because it corresponds to the limit $p_F \to 0$, and the number density goes to zero in this limit, $n \propto p_F^3 \to 0$. The validity of the results of Eqs. (9.22) and (9.23) for this idealised model must be checked by using the exact expression for the response tensor for a completely degenerate relativistic electron gas. The relevant response tensor is that derived by Jancovici, cf. Eq. (9.16). We emphasise that it is essential to use the relativistic result, rather than the more familiar nonrelativistic theory due to Lindhard [7], which does not include dispersion due to PC. We looked for both longitudinal and transverse pair modes but found no solutions when the exact Jancovici response functions were used.

We conclude that the idealised model for a degenerate electron gas incorrectly predicts pair modes whereas no pair modes exist in an actual electron gas with $p_F \neq 0$. Whereas the response tensor, Eq. (9.8), with the numerator of the form Eq. (9.10), has poles at $\omega^2 = \omega_\pm^2$, cf. Eq. (9.18) for a cold distribution, the correct expression for the response tensor has no pole. Specifically, for any $p_F \neq 0$ the logarithmic function in Eq. (9.16) has no singularity at the threshold for PC. In contrast, in a degenerate Bose gas there is a pole due to the contribution from particles in the ground state (Bose condensate). This suggests that pair modes are present only in Bose gases below their degeneracy temperature, because it is only in such cases that the relevant pole appears in the response function.

9.4.6. *Wave properties for a spin 1 gas*

For a completely degenerate, unpolarised spin 1 plasma, the longitudinal dispersion equation is the same as in the spin 0 case [2], (see the coefficients in Table 9.1). The transverse dispersion equation is

$$\omega^6 - \omega^4 \left(3|\mathbf{k}|^2 + \omega_p^2 + 4m^2\right)$$
$$+ \omega^2 \left(3|\mathbf{k}|^4 + 2\omega_p^2|\mathbf{k}|^2 + 4|\mathbf{k}|^2 m^2 + 4\omega_p^2 m^2\right) - |\mathbf{k}|^4 \omega_p^2 - |\mathbf{k}|^6 = 0. \quad (9.24)$$

For $\omega_p^2 \ll m^2$ the three solutions may be approximated by

$$\omega^2 \approx 4m^2 + 2|\mathbf{k}|^2, \qquad \omega^2 \approx \omega_p^2 + |\mathbf{k}|^2, \qquad \omega^2 \approx \frac{|\mathbf{k}|^4}{4m^2}. \quad (9.25)$$

The second of these is the familiar transverse mode, and the modifications to it due to PC are unimportant with the approximations made here. The first mode is a pair mode and third is a roton-like mode. At cutoff ($|\mathbf{k}|^2 = 0$)

the frequency of the pair mode is equal to the frequency, Eq. (9.18) defined by the zero of the resonant denominator, Eq. (9.3), and it is greater than the frequency, Eq. (9.18), for $|\mathbf{k}|^2 > 0$. Like the longitudinal pair mode, it is in the regime where PC is allowed.

9.5. Properties of pair modes

Besides the dispersion relation and the polarization vector, other characteristic properties of waves in a given mode relate to the distribution energy between the electromagnetic field and particle motions associated with the wave, and to the propagation of wave energy. These properties are discussed here for pair modes.

9.5.1. *Wave energetics*

The properties of an arbitrary wave mode, labelled M, include the dispersion relation, $\omega = \omega_M(\mathbf{k})$, the polarization vector, $\mathbf{e}_M(\mathbf{k})$, and the ratio of electric to total energy in the waves, $R_M(\mathbf{k})$ [8]. For the pair mode in a completely degenerate, spin 0 gas, the dispersion relation is given by Eq. (9.23), with $M \to +$, and the polarization is longitudinal, $\mathbf{e}_M(\mathbf{k}) \to \mathbf{k}/|\mathbf{k}|$. In this case, the only energies are the electric energy and the particle energy associated with the perturbed motion of particles in the wave. (There is no magnetic energy in a longitudinal wave.)

This ratio of electric to total energy in a longitudinal ($M \to L$) wave is given by [8]

$$R_L(\mathbf{k}) = \left(\left\{ \frac{\partial}{\partial \omega} \left[\omega K^L(k) \right] \right\}_{K^L(k)=0} \right)^{-1}, \tag{9.26}$$

where $K^L(k) = 1 - \mu_0 \mathcal{P}^L(k)/\omega^2$ is the longitudinal dielectric constant (in the rest frame of the plasma). For the pair mode in a completely degenerate, spin 0 gas, using the approximations made in Eq. (9.20), with $M \to +$, Eq. (9.26) may be approximated by

$$R_L(\mathbf{k}) \approx \frac{|\mathbf{k}|^2}{8m^2}. \tag{9.27}$$

It follows that, for $|\mathbf{k}|^2 \ll m^2$, only a small fraction of the energy is associated with the electric field, and most of the wave energy is associated with the perturbed motion of the particles. One may ascribe a total energy $\omega \approx 2m$ and a momentum $|\mathbf{k}|$ to a quantum of a pair mode, and then Eq. (9.27) implies an electric energy $|\mathbf{k}|^2/4m$ associated with the wave quantum. A physical model for the wave is needed to interpret this result, and no suitable model is available.

The wave energetics also includes the energy flux, which is proportional to the energy density times the group velocity. In the case of a spin 0 plasma at absolute zero, the energy flux in the pair mode is due entirely to the kinetic energy flux (there is no Poynting flux in a longitudinal wave). The group velocity, $\mathbf{v}_{gM} = \partial \omega_M(\mathbf{k})/\partial \mathbf{k}$, for the pair mode, follows from Eq. (9.26) with $M \to +$, and may be approximated by

$$\mathbf{v}_g \approx 2\mathbf{k}/(4m^2 + |\mathbf{k}|^2)^{1/2}. \tag{9.28}$$

The group velocity is interpreted as the velocity of energy propagation, and the phase velocity is the inverse of the ratio of the wave momentum to the wave energy, $\mathbf{k}/\omega$. In this case the product of these two velocities is equal to 2, but the significance (if any) of this is unclear to us.

9.5.2. *Generation and damping of pair modes*

Pair mode waves have phase speed greater than the speed of light, and hence cannot be generated through Cerenkov emission. The pair mode, Eq. (9.21), in a degenerate spin 0 gas has a frequency in the regime where PC is possible. Hence these waves may be created and damped through pair annihilation and creation. Once generated, pair modes can play analogous roles to other waves in plasmas. In particular, they can be involved in three-wave interactions, with the decay of a pair mode wave into two ordinary transverse waves being possible.

A related point concerns a thermal distribution of pair modes. When relativistic quantum effects are taken into account, a gas in thermal equilibrium at a finite temperature contains both particles and antiparticles, and it also contains a thermal distributions of all waves in all wave modes, including pair modes. The energy density in any wave mode is

$$W_M = \int \frac{\omega_M(\mathbf{k})}{\exp[\omega_M(\mathbf{k})/T] - 1} \frac{d^3\mathbf{k}}{(2\pi)^3}, \tag{9.29}$$

where T is the temperature. With $\omega_M(\mathbf{k}) \sim 2m$, the integrand is extremely small for nonrelativistic temperatures, $T \ll m$.

9.5.3. *Super dense plasmas*

In their discussion of pair modes, Kowalenko, Frankel and Hines [1] considered the case of a super dense plasma, with $\omega_p^2 > m^2$. Discussion of this limit in connection with dispersion in an electron gas has been criticised [9]. The point criticised is the claim that for $\omega_p^2 > 4m^2$ the dispersion curves for longitudinal and transverse waves enter the region $\omega^2 = 4m^2 + |\mathbf{k}|^2$ where PC occurs [10, 11]. But the criticism contains an inconsistency, namely that the term $(k^2/2m)^2$ in the resonant denominator was neglected [9] in discussing the limit $|\mathbf{k}| \to 0$. The neglect of the term $(k^2/2m)^2$ corresponds to neglecting dispersion due to PC, and this is unjustifiable in the context. Nevertheless, the limit $\omega_p^2 > m^2$ needs to be treated with more care than it has hitherto. In particular, at such extreme densities, macroscopic mass renormalization should cause the properties of the particles to be substantially different from those of free particles in vacuo. The results in the present paper should be valid for $\omega_p^2 \ll m^2$, but their extension into the regime $\omega_p^2 \gtrsim m^2$ is questionable.

9.6. Discussion and conclusions

The original identification of a pair mode [1] was as a longitudinal wave in a completely degenerate gas of bosons with spin 0. Obvious questions are whether there are corresponding transverse pair modes in Bose gases, and whether analogous modes exist for an electron gas, that is, for fermions rather than bosons. These questions are answered here for completely degenerate gases: there is no transverse pair mode in a spin 0 gas, there are no pair modes at all in an electron gas, and there are both longitudinal and transverse pair modes in a spin 1 gas. These results are derived here only for completely degenerate gases.

Is there a simple physical model that describes the properties of pair modes? In the cases we consider, the pair modes exist only in the regime where PC (pair creation and annihilation) is allowed. This suggests a physical model of a quasi-bound pair that can decay into a real pair. A pair mode has some similarity to a collection of virtual pairs modified ('dressed' in conventional jargon) by the dispersion of the plasma, with the relevant dispersion being that associated with PC. However, the role of the electromagnetic field is unclear. Coulomb coupling between a particle and antiparticle can play no role in such a physical model: pair modes appear as solutions of the linear wave equation and particle-particle interaction are not taken into account in deriving the linear response. Also, the fact that a transverse pair mode exists in a degenerate spin 1 plasma implies that the role of the electromagnetic field in the quasi-pairs does not depend on it being longitudinal. The qualitative picture that pair modes act like virtual pairs is supported by our estimate of the ratio of electric to total energy in the waves, cf. Eq. (9.27), which shows that the energy is mainly in the particles (for $|\mathbf{k}|^2 \ll m^2$), and may be plausibly interpreted in terms of the rest energy of a (virtual) pair. No simple physical model for a pair mode is available, and their intrinsic dependence on dispersion due to PC makes it difficult to identify an appropriate analogy. One possible analogy is with creation and annihilation of electrons and holes in semi-conductors, but we have been unable to find any discussion of a counterpart of pair modes for such a system.

Our failure to find pair modes in a completely degenerate electron gas suggests that pair modes are intrinsic to degenerate boson plasmas. The argument for this is that pair modes are associated with a pole in the response function. The dispersion associated with this pole leads to the characteristic dispersion relations, Eq. (9.18), for pair modes and roton-like modes. However, only particles in the ground state (in a Bose condensate) lead to such a pole. For a completely degenerate electron gas there is no pole, and the dispersion is determined by logarithmic contributions, specifically those that appear in Jancovici's response function, Eq. (9.16). In a conventional treatment of a Bose gas below its degeneracy temperature, there are contributions from the particles in the ground state and in excited states that are treated as a continuum [6]. The ground state contribution to the response tensor has a pole, as in the completely degenerate case. The continuum of excited states gives a contribution that involves only logarithmic contributions, similar to Jancovici's response function. This leads us to conclude that pair modes exist only below the degeneracy temperature in Bose gases. Above the degeneracy temperature only the continuum of excited states contributes, and there is no pole in the response tensor. In brief, pair modes and roton-like modes appear to be associated only with the Bose condensate in a Bose gas below its degeneracy temperature.

In conclusion, the identification of pair modes by Hines and collaborators [1] raises interesting questions concerning the existence and significance of these modes. In this paper we identify these modes as characteristic of a degenerate Bose gas. A simple physical model for pair modes remains elusive.

Acknowledgments

We thank Jeanette Weise and Qinghuan Luo for helpful comments on the manuscript.

References

[1] V. Kowalenko, N. F. Frankel, and K. C. Hines, *Phys. Rep.* **126**, 110 (1985).

[2] D. R. M. Williams and D. B. Melrose, *Aust. J. Phys.* **42**, 59 (1989).

[3] L. M. Hayes and D. B. Melrose, *Aust. J. Phys.* **37**, 615 (1984).

[4] B. Jancovici, *Nuovo Cim.* **25**, 428 (1962).

[5] H. D. Sivak, *Ann. Phys.* **159**, 351 (1985).

[6] L.D. Landau and E.M. Lifshitz, *Statistical Physics*, Pergamon Press (1959).

[7] D. J. Lindhard, *Mat. Fys. Medd. Dan. Vid. Selsk.* **28**, 1 (1954).

[8] D. B. Melrose, *Instabilities in space and laboratory plasmas*, Cambridge University Press, (1986)

[9] E. Braaten and D. Segel, *Phys. Rev. D* **48**, 1478 (1993).

[10] V. N. Tsytovich, *Sov. Phys. JETP* **13**, 1249 (1961).

[11] D. B. Melrose and L. M. Hayes, *Aust. J. Phys.* **37**, 639 (1984).

Chapter 10

Ultrahigh Energy Cosmic Rays

Bruce H. J. McKellar

School of Physics, University of Melbourne, Victoria, Australia 3010
Email: bhjm@unimelb.edu.au

A review of the experimental observations of ultrahigh energy cosmic Rays as in 2005 is given, together with theoretical proposals for understanding those observations. This follows the paper given at the mini-conference commemorating Ken HInes. This is then updated to the present day, taking account of the recent results from the Pierre Auger Observatory.

10.1. Introduction

This paper is primarily based on the talk I gave at the mini-conference commemorating Ken Hines. For historical accuracy, I have not modified the primary portion of the paper, simply updating it in the final section. In that section we will see that, perhaps unfortunately, recent results from the Pierre Auger Observatory in Argentina show that the ultrahigh energy cosmic rays are not over abundant, and there may not be any new physics at work.

At the mini-conference I chose to discuss cosmic ray physics because

- Ken's early post-graduate work was on cosmic rays, and his supervisor was Henry Rathegeber, who later taught me at Sydney University.
- In the years before the mini-conference, my collaborators and I had been working on the subject of neutrino clustering, and in the influence of that clustering on a model for the generation of ultrahigh energy cosmic rays [1, 2]. This work, done with Dr Matt Garbutt, now with the Department of Defence, Canberra; Dr Terry Goldman, Los Alamos National Laboratory, and Professor Jerry Stephenson, University of New Mexico, is the basis of the work reported at the mini-conference and in this Chapter.

10.2. History

The highest energy cosmic ray events are studied by the extensive air showers which they create, and the energy of the air shower is determined from the density distribution of the air-shower particles on the earth's surface by fitting the observed density to the Nishimura, Kamata, and Greisen lateral distribution function [3]. Larger energy showers occur less frequently, so a large collecting area is required to see them with reasonable frequency and to determine the energy. One of the events seen at the first large air-shower detector in 1963 at Volcano Ranch [4] in New Mexico had an energy in excess of 10^{20} eV, and 6 had energies above $10^{19.6}$ eV.

In 1965, the cosmic microwave background radiation was discovered [5], and almost immediately, Greisen [6] and Zatsepin and Kuzmin [7] pointed out that collisions of high energy protons with the cosmic microwave photons became inelastic for proton energies above $\sim 10^{19.3}$ eV, which is above the threshold for $\gamma_{2.7K} + p \to \Delta \to \pi + N$. An initial proton beam above that energy will be attenuated with a mean free path which is 3 to 7 Mpc [8] for proton energies above 10^{20} eV. It is hard enough to find a mechanism which will produce protons of an energy greater than 10^{20} eV, let alone so much larger that is can be attenuated down to those energies in traversing the CMB. It is generally agreed that for a proton with energy above the threshold to be observed at earth it must be produced within a sphere or radius not larger than ≈ 100 Mpc. Recent calculations favour a lower limit, about 50 Mpc [9]. Greisen said of the Volcano ranch event: "even the one event recorded at 10^{20} eV appears surprising".

Because of the sharp cutoff in the Planck distribution of the photons, the spectrum of incoming protons will be strongly attenuated at about $10^{19.5}$ eV, unless they are produced within about 50 Mpc of the earth. This is referred to as the GZK cutoff, and cosmic rays with energies above this limit are referred to as ultrahigh energy cosmic rays or UHECRs. Indeed a similar sharp energy cutoff occurs for nuclei and gammas as well. Only neutral primary particles which also lack strong interactions, such as neutrinos, can evade the limit.

Because measuring the lateral density distribution of the air-shower particles is is susceptible to fluctuations depending on the stage of shower development i.e. the depth in the atmosphere at which the initial interaction took place, and thus to calibration errors, it is useful to have an alternative method of determining the energy in the shower, and thus the energy of the primary particle. An alternative method is to collect the fluorescence light resulting from the excitation of nitrogen molecules by ionising radiation within the shower. This requires a system of telescopes and photomultipliers observing the atmosphere. It was implemented in the Fly's Eye experiment in Utah, which was later improved by duplication of the detector to give a stereoscopic view of the atmosphere, giving the HiRes experiment. The techniques was then incorporated into the Pierre Auger Observatory, which combines a ground array and the atmospheric techniques.

10.3. The UHE data

The highest energy CR event was measured by the Fly's Eye experiment at $E_0 = (3.2^{+0.36}_{-0.54}) \times 10^{20}$ eV, followed by an event recorded by the Akino Giant Air Shower Array (AGASA) in Japan of $E_0 = (2.10 \pm 0.5) \times 10^{20}$ eV. There is still some debate as to the exact location of the GZK cutoff, however the consensus is that it should be seen below $E_0 = 10^{20}$ eV, and now there are about 30 events above this energy. However cosmic rays with energies above 10^{19} eV (= 10 EeV) have a very low flux, about 1 per square Kilometre per second, so the statistics are poor and there is a lot of scatter in the data. Bahcall and Waxman [10] have pointed out that all of the data except the AGASA results are in fact consistent with the GZK cutoff, and that one can renormalize the AGASA results to restore their consistency. As I remember from my student days at Sydney University, air-shower results in cosmic ray physics can be controversial.

If we take the above GZK events at face value, they pose two questions.

(1) By what mechanism are particles being accelerated to post GZK energies?
(2) How do they arrive at earth with these energies?

The first question will be dealt with in Sec. 10.4. An obvious solution to the latter question is that the sources be distributed locally, however this has all but been ruled out by studies of anisotropy in the UHE data. After discussing this aspect of the data, I return to the second question in Sec. 10.5.

10.3.1. *Isotropy*

For primary particles with energy below about 2×10^{17} eV originating within the galaxy, an isotropic spectrum is expected as the particles diffuse through the irregular galactic magnetic field. However for energies above

2×10^{17} eV, the effects of the magnetic field become less important and the trajectories of the primary particles should point back to their sources. If the sources of UHECRs are distributed in the same way as other matter, i.e. in the galactic plane, then an anisotropy in that direction should be apparent. This argument assumes that the galactic magnetic field is not very much greater than currently thought. It should also be noted that primary particles with a large charge (such as iron nuclei) would suffer greater deflections in the galactic and inter-galactic field.

While there is some evidence of anisotropy for primary energies of about $E_0 = 10^{18}$ eV, the UHE data does not display any anisotropy on a large scale, indicating that the post-GZK CRs are of extra galactic origin. There is however some indication of a small scale anisotropy, possibly indicating production by point sources.

10.4. Acceleration methods

The process by which particles may be accelerated to post-GZK energies is as much a mystery as their propagation to earth. The two conventional methods for accelerating charged particles are direct acceleration in an electromagnetic field and statistical acceleration at a number of acceleration points. The latter is known as Fermi acceleration. In any case, the observational constraints on acceleration mechanisms are the observed power law governing the CR energy spectrum, and limits on the fluxes of γ-rays and neutrinos at lower energies resulting from EM cascades caused by the accelerated particle. Direct acceleration is disfavoured since it is unable to reproduce the observed power law, while both mechanisms have their parameter space stretched in trying to achieve post-GZK energies.

The most efficient form of statistical acceleration is called 1st order Fermi-acceleration, and occurs at the front of shock waves resulting from supernovas or other energetic astrophysical events. The simplest theory models the shock wave as a non-relativistic transverse wave with its associated magnetic field normal to the shock front. The CRs scatter off inhomogeneities in the magnetic field. Should they cross the shock front several times they can be accelerated to very high energies.

The model of shock acceleration results in the following limits on the maximum energy attainable [11]:

$$E_{\text{max}} \leq 6.0 \times 10^2 \xi^{1/2} i \left(\frac{B}{1T} \right)^{-1/2} \qquad \text{GeV electrons,} \quad (10.1)$$

$$\leq 2.0 \times 10^9 \xi^{1/2} \left(\frac{B}{1T} \right)^{-1/2} \qquad \text{GeV protons,} \qquad (10.2)$$

$$\leq 2.0 \times 10^9 \xi^{1/2} \frac{A^2}{Z^{3/2}} \left(\frac{B}{1T} \right)^{-1/2} \qquad \text{GeV nuclei,} \qquad (10.3)$$

where $\xi \leq 1$ accounting for energy loss processes and B is the strength of the magnetic field.

A dimensional analysis given by Hillas [12] allows one to place a bound on the maximum attainable energy which is independent of the acceleration mechanism:

$$E_{\max} \leq \frac{1}{2} Z \beta \left(\frac{R}{\text{kpc}} \right) \left(\frac{B}{\mu \text{G}} \right) \text{EeV}, \qquad (10.4)$$

where R is the size of the accelerating region with an associated magnetic field B. The assumption here is that whatever the accelerating mechanism a particle will make many loops in the accelerating region, which must therefore be larger than twice the Larmor radius (cyclotron radius) of the particle lest it be lost from the region. Hillas found that very few astrophysical objects have either the high magnetic field or the very large accelerating region required to confine the particle.

Hillas' original list of potential sources of post-GZK CRs included neutron stars, white dwarfs, Active Galactic Nuclei (AGN), Radio Galaxy lobes (RG) and acceleration in an extended galactic halo or galactic cluster. Since then Gamma Ray Bursts (GRBs) and colliding galaxies have been added to the list of possible sources [13].

10.5. How do post GZK particles get here?

The conventional acceleration mechanisms discussed in the previous section are all theoretically capable of accelerating particles up to $E_0 = 10^{20}$ eV. For charged particle primaries to avoid the GZK cutoff they must originate within $\simeq 50$ Mpc, but in doing so the trajectories of the CR should point back to the sources thus violating the observed large scale isotropy. Thus alternative, unconventional sources of the primary cosmic rays have been considered.

10.5.1. *Top-down models*

One proposal is that the UHECRs may be a result of decay of hitherto unknown supermassive particles known generically as X-particles. The ensemble of theories describing various mechanisms by which the X-particles are produced and decay have become known as 'Top-down' solutions to the UHECR puzzle. There are a plethora of top-down theories, most of which predict that the X-particles were produced in the big bang and have a mass of the order of the Planck mass, $M_p \simeq 10^{25}$ eV. The decay products of such a massive particle would have ample energy to propagate cosmological distances and still arrive at earth with post-GZK energies [14].

10.5.2. *Violations of fundamental laws*

A different scenario proffered as an explanation to the non-observation of the GZK cutoff is the violation of Lorentz symmetry at very high energies [15, 16]. In some string theories it can be shown that energy is not conserved exactly, rather it is only conserved on average or statistically [15]. This has the effect of raising the threshold energy for photopion production in the case of a proton propagation, and thus moving the GZK cutoff to a higher energy, with similar results for Nuclei and photon propagation. If this were the case one could imagine the GZK events originating from conventional sources distributed at extra-galactic distances.

10.5.3. *The Z-Burst model—UHE neutrinos*

Another interpretation of the UHECR data and the one which is the focus of this section is known as the Z-Burst model [17, 18]. In this scenario UHE neutrinos annihilate at the Z resonance on Cosmic Background Neutrinos (CBNs) which are relics from the big bang. The resulting resonant Z primarily decays into a jet of hadrons, a fraction of which produce particles with energy greater than the GZK threshold. So long as the annihilation occurs within a radius of $R_{GZK} \simeq 50$ Mpc, some of the decay products will be observed as UHECRs. This scenario is attractive as it avoids the complications of the GZK cutoff since the neutrino is neutral and weakly interacting. But it does not introduce new physics.

In the simplest Z-Burst model, with the relic neutrino density is taken to be $n_\nu = 54$ cm^{-3} and $D = 50 - 100$ Mpc, the flux of UHECRs has been computed and compared to the observed fluxes, which are $F_{p/\gamma} = 10^{-19}$ cm^{-2}s^{-1}str^{-1} above 5×10^{19} eV and $F_{p/\gamma} \sim 2 \times 10^{-20}$ cm^{-2}s^{-1}str^{-1} above 5×10^{20} eV [17].

While the Z-Burst model provides an economical way around the GZK cutoff, it relies on there being a significant cosmic neutrino flux or in the absence of this an enhancement of the relic neutrino background over the standard theory. The means by which a flux of neutrinos at energy $E_\nu \sim 10^{21}$ eV is produced, as is required should the relic neutrinos possess a mass in the eV range, becomes an issue. Not surprisingly the actual neutrino flux has not been measured at this energy, and one must rely on models of the cosmic neutrino flux to access the viability of the Z-Burst mechanism. In principle, one can get the neutrino flux by acceleration of protons to sufficiently high energies, or by decay of some X-particle. As we want to reduce the new physics involved in our discussion we will not consider the latter further.

In order to resonantly annihilate on relic neutrinos with mass of $m_\nu \leq 1$ eV, cosmic neutrinos need an energy of $E_\nu \geq 10^{21}$ eV. This energy requirement pushes the parameter space of conventional astrophysical sources to the limit, since the main mechanism for production of UHE neutrinos is through the decay of charged pions as a result of protons scattering resonantly off photons, the required energy of the primary nucleon would be $E_0 \simeq 10^{23}$ eV; an extraordinarily high value. Nonetheless simulations of AGNs and GRBs predict a small neutrino flux at this energy. Mannheim, Protheroe and Rachen performed a calculation that took into account energy losses due to photopion production and red shifting of sources [19]. Waxman and Bahcall [20] have performed similar calculations to that of Mannheim *et. al.*, with a different normalisation of the spectrum. They obtained an upper bound of the neutrino flux which is a factor of 3 less than the estimate of Mannheim *et al.*

The predicted neutrino flux is also dependent on the background neutrino density, which is the standard cosmological model is $n_\nu = 54$ cm^{-3} per neutrino flavour. With this minimal density and a GZK zone of $D = 50$ Mpc, the model predicts a flux of UHECRs much less than that observed. If one retains the standard background neutrino density, then the neutrino flux must be increased, with the result that the energy density of the UHE neutrinos is much larger than the energy density of the observed CR spectrum, or indeed that of the universe. So we turn to a way of increasing the density of background neutrinos.

10.6. Neutrino clouds

In a paper by Stephenson, Goldman, and McKellar [21], a mechanism was developed by which the relic neutrinos would condense to form high density neutrino clouds. Such a neutrino cloud of appropriate dimensions would provide a much denser target for the cosmic ray neutrinos. Neutrino clouds may form if a very light scalar particle couples to neutrinos only. A variant of the standard model in which this occurs has been given by He, McKellar, and Stephenson [22].

At a phenomenological level, the Lagrangian density for a neutrino field ψ interacting with a scalar ϕ is:

$$\mathcal{L} = \bar{\psi}(i\partial_\nu)\psi + \frac{1}{2}[\phi(\partial^2 - m_s^2)\phi] + g\bar{\psi}\psi, \qquad (10.5)$$

which results in a coupled Dirac and Klein-Gordon equation:

$$(\partial^2 + m_s^2)\phi = g\bar{\psi}\psi, \qquad (10.6)$$

$$(i\partial\!\!\!/ - m_\nu)\psi = -g\phi\psi. \qquad (10.7)$$

The phenomena that leads to the formation of neutrino clouds may be illustrated by considering the solutions to these equations for the static and rotationally invariant case. The non-homogeneous Klein-Gordon equation gives for the scalar field:

$$\phi = \frac{g}{m_s^2}\bar{\psi}\psi, \qquad (10.8)$$

which can be substituted into the Dirac type equation to define an effective, reduced neutrino mass m_ν^*:

$$m_\nu^* = m_\nu - \frac{g^2}{m_s^2}\bar{\psi}\psi, \qquad (10.9)$$

when the neutrino density is large, such as it was during the time just after the Big Bang the operator $\bar{\psi}\psi$ may be treated statistically and evaluated as an expectation value:

$$<\bar{\psi}\psi> = \frac{\gamma}{(2\pi)^3}\int_0^{k_F} k^2 dk \frac{m_\nu^*}{\sqrt{(m_\nu^*)^2 + k^2}}, \qquad (10.10)$$

where γ is the neutrino degeneracy while k and k_F are the neutrino momentum and Fermi momentum respectively and so,

$$m_\nu^* = m_\nu - \frac{g^2\gamma}{2\pi^2 m_s^2}\int_0^{k_F} k^2 dk \frac{m_\nu^*}{\sqrt{(m_\nu^*)^2 + k^2}}. \qquad (10.11)$$

The difference between the effective neutrino mass and the vacuum mass represents a binding energy, hence work must be done to move a neutrino from a region of high to low density. Under certain conditions this binding energy can facilitate the formation of clouds as the natural tendency is now for neutrinos to cluster. It is important to note that the formation of clouds in this scenario requires that the already bound relic neutrino gas to break up, rather than relying on trapping of free neutrinos. Thus it avoids the Tremaine-Gunn bound on the density of neutrino clusters.

The transition from infinite to finite matter is made by studying the equations of motion using the Thomas Fermi approximation:

$$\nabla^2 \phi + m_s^2 \phi = g < \bar{\psi}\psi >, \qquad (10.12)$$

where the assumption that the clouds are static is retained. The strongest experimental constraint comes from an argument involving the observation of neutrinos from SN1987A. By requiring that the cross section for scalar exchange between neutrinos be sufficiently small so as not to effect the spectrum of supernova neutrinos, an upper bound on the coupling strength is found:

$$\tilde{\alpha} = g^2/4\pi < 10^{-9}. \qquad (10.13)$$

We consider clouds of density $10^{10} - 10^{14}$ cm^{-3} with radii in the range $10^{14} - 10^{20}$ cm.

First we establish whether or not a neutrino cloud can alleviate the problems associated with the Z-Burst Model. I then take the reverse point of view and ask if the UHECR data can place any bounds on the size and density of a neutrino cloud.

A feature of the generic Z-burst model is that it allows a minimum required neutrino flux to be defined. If it is accepted that the UHE CRs are produced by this mechanism and assumed that the neutrino relic density is such that the neutrino flux originating from the source is resonantly absorbed in its entirety, this flux establishes the minimum required, which is $5.6 \times 10^{-21} \leq E_R F_{\nu\ min}(E_R, 0) \leq 5.6 \times 10^{-20}$ cm^{-2}s^{-1}str^{-1} depending on the experimental value of $F_{p/\gamma}$.

To determine the flux of post GZK events the multiplicity (M), and the respective momentum of the decay products of the resonant Z, are needed. In the case of neutrino clouds the multiplicity has added importance since, to the authors knowledge, the data never shows post-GZK events in coincidence, hence the M decay products must be spread over an area such that the various experiments designed to detect these particles never see more than one at a time. The decay products are boosted, in the reference frame of the earth, into a cone of opening angle $\theta = 1/\gamma_Z = 2(m_\nu/eV) \times 10^{-11}$ radians [17, 18]. Each decay product will on average occupy an area of approximately:

$$A_p = \frac{\pi}{M}(r_{int}\tan\theta)^2 \approx \pi \times 10^{-22}\frac{r_{int}^2}{M},$$

where r_{int} is the distance from earth at which the annihilation takes place. The requirement that no coincident events be observed is met if the effective detection area A_{det} of the experiment is less than A_p.

The effective detection area is defined by the particular detector's acceptance (sometimes called the aperture) $\mathcal{A}$,

$$\mathcal{A} = A_{\det}\int_0^{\theta_{\max}}\cos\theta d\Omega \tag{10.14}$$

where $\theta_{\max}$ is the maximum zenith angle for the particular event energy. The AGASA and Fly's Eye experiments both report apertures of ~ 1000 km^{2}sr with an effective detection area of several hundred square kilometres. Hence A_p is constrained to be larger than say ~ 500 km^2.

A detailed discussion of the spectrum of particles resulting from the Z-burst is given by Weiler, based on the results from LEP. The mean multiplicity for decay to hadrons is 40, based on a hadronic branching ratio of 70%, a neutrino branching ratio of 20% and a charged lepton branching ratio of 10%. This results in 15 π^0's, 28 $\pi^\pm$'s and 2.7 nucleons after the decay of various short lived mesons. The charged pions will then decay into three neutrinos and an electron or positron within approximately 0.02pc. The dominant decay mode of the π^0's is to two photons, this occurs over tens of kilometres. The multiplicity of cosmic rays reaching the earth from one Z-burst is thus ~ 60 on average, where the neutrinos resulting from π-decay have been discounted. Of these sixty particles only thirty or so are expected to have energies above the GZK cutoff, so the multiplicity for post GZK events would be $M \sim 30$.

With a multiplicity of ~ 60 and a detection area of $A_{\det} \sim 500$ km^2 a sphere of radius $R_c \sim 10^{18}$ cm can be defined within which neutrino annihilations will on average result in the observation of a coincident event. The rate of coincident events, $F_{coincident}$, will depend on the neutrino flux incident upon this sphere, and can be readily evaluated. If in addition one makes the simplifying assumption that the relic neutrino column density outside the cloud is much less than the column density inside, the flux of coincident events becomes.

$$F_{\text{coincident}} = E_R F(E_R, 0) \int \frac{ds}{M_Z^2} \exp[-\sigma(s) n_\nu R]$$

$$\times \left(\exp[-\sigma(s) n_\nu R_c] - 1 \right). \qquad (10.15)$$

The non-observation of a coincident flux implies the following bound,

$$\mathcal{A} \, t_r F'_{\text{coincident}} < 1, \qquad (10.16)$$

where again $\mathcal{A}$ is the experimental aperture at $E_0 > 10^{19.6}$ eV and t_r is the experimental running time.

For the neutrino cloud hypothesis to hold true, values of cloud density and radius must be found so that for each experiment,

$$F'_{\text{coincident}} \geq E_R F(E_R, 0) \int \frac{ds}{M_Z^2} \exp[-\sigma(s) n_\nu R] \left(1 - \exp[-\sigma(s) n_\nu R_c] \right). \qquad (10.17)$$

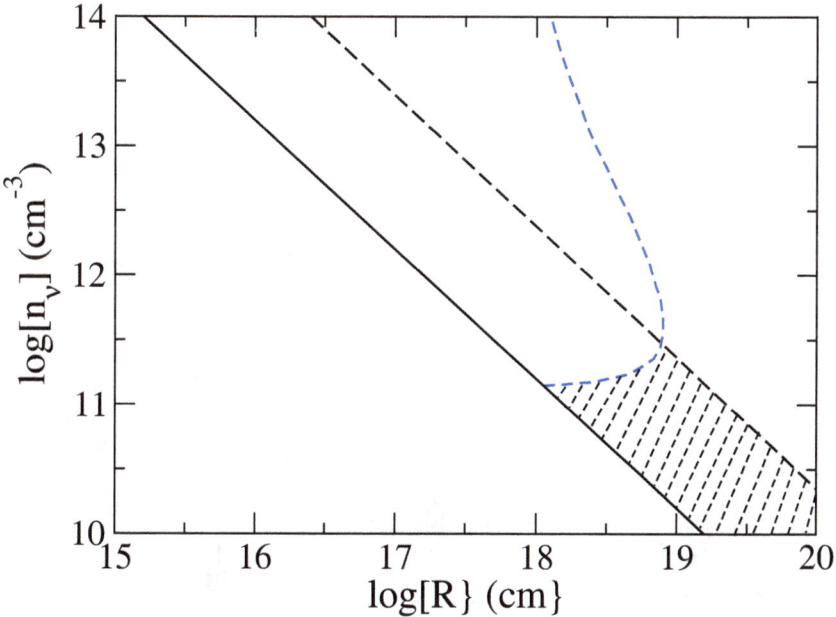

Fig. 10.1. The solid line represents the cloud parameters required to produce a GZK flux of $F_{p/\gamma} = 10^{-20}$ cm^{-2}s^{-1}str^{-1}, with an incident neutrino flux of $E_R F_\nu(E_R, 0) = 10^{-19}$ cm^{-2}s^{-1}str^{-1}. The long-dashed line represents a GZK flux of $F_{p/\gamma} = 10^{-19}$ cm^{-2}s^{-1}str^{-1}.

The AGASA experiment currently has the largest exposure quoted at:

$$\mathcal{A}t_r = 670 \text{ km}^2\text{sr yr} \simeq 2 \times 10^{21} \text{ cm}^2 \text{ sr s}. \tag{10.18}$$

This allows a coincident flux of $F'_{\text{coincident}} \simeq 5 \times 10^{-22} \text{ cm}^{-2}\text{s}^{-1}\text{str}^{-1}$. The values of R and n_ν which solve Eq. (10.17) are shown in Figs. 10.1 and 10.2. The short-dashed line in each is the bound on R resulting from the non-observation of coincident events. The hashed region in both figures denotes the allowed parameter space. The coincidence constraint allows a minimum cloud radius of $R_{\text{min}} = 10^{18}$ cm to be specified, and in some sense a maximum relic neutrino density, depending on the incident neutrino flux.

The constraints on the parameter range of neutrino clouds as a result of the arguments based on coincident events have not taken into account the spatial resolution of the various CR experiments. While this is not an issue for medium to large clouds, it does become important for clouds with smaller radii. For example, if a neutrino cloud had a radius of 50 au the decay products of a Z-burst would be confined to a cone of radius less than 10 m at earth. This radius would be too small for most CR experiments to

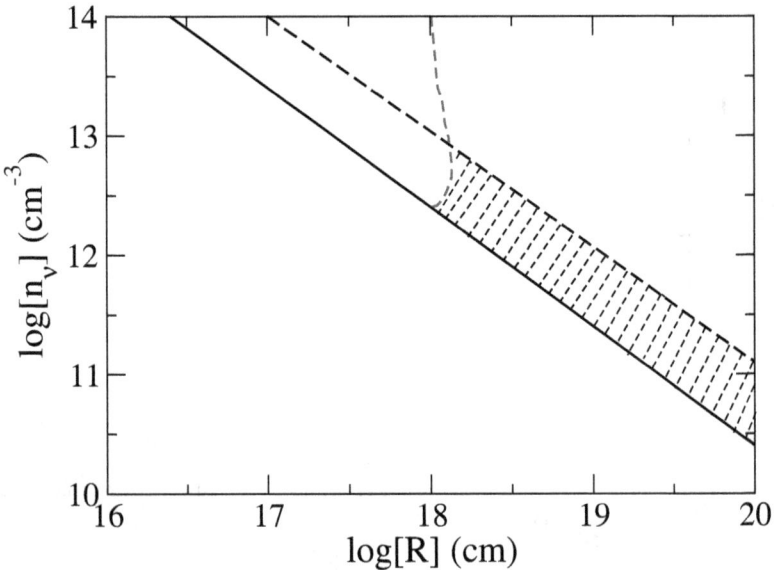

Fig. 10.2. The solid line represents the cloud parameters required to produce a GZK flux of $F_{p/\gamma} = 10^{-20} \text{ cm}^{-2}\text{s}^{-1}\text{str}^{-1}$, with an incident neutrino flux of $E_R F_\nu(E_R, 0) = 10^{-20} \text{ cm}^{-2}\text{s}^{-1}\text{str}^{-1}$. The long-dashed line represents a GZK flux of $F_{p/\gamma} = 1.7 \times 10^{-20} \text{ cm}^{-2}\text{s}^{-1}\text{str}^{-1}$, the maximum flux for this small incident neutrino flux.

resolve, in which case the decay products will be measured as a single event with an energy of E_R. In the minimal Z-burst model E_R is a significantly higher energy than the highest energy CR event. However for high density clouds the effect of the Fermi motion is to reduce the average value of E_R so that the possibility of small radii clouds is not excluded.

10.7. Conclusion

If the data continues to favour a flux of ultrahigh energy cosmic rays above the GZK limit which exceeds the flux expected from attenuation in the cosmic microwave background, we are in for an exciting time in physics, because we will need to invoke physics beyond the standard model to explain the data.

Many models have been proposed. The model of neutrino clouds goes beyond the standard model by introducing a very light scalar particle interacting very weakly with neutrinos. It provides a way of enhancing the density of neutrinos in a region relatively near the earth and allows the Z burst mechanism for generating ultrahigh energy cosmic rays to give a flux of UHECRs consistent with experiments, with an incident neutrino flux consistent with expectations. The model is in turn heavily constrained by the data.

10.8. Epilogue

Recent results from the Pierre Auger Observatory [23] show that the flux of UHECRs above 4×10^{19} eV steepens significantly. The spectral index in the decade below 4×10^{19} eV is 2.69 ± 0.06 (I have added the statistical and systematic errors in quadrature, but the latter dominate), while that above 4×10^{19} eV is $4.2 \pm .4$ (this time the statistical errors dominate). This is consistent with the GZK cutoff. Moreover they have shown that there is a reasonable correlation between the arrival directions and Active Galactic Nuclei within 75 Mpc of earth [24], and that less than 10% of the showers above 10^{19} eV have photon primaries [25]. These new data are consistent with the hypothesis that the ultrahigh energy cosmic rays are completely normal, and that no new physics is necessary to explain them. While the photon data is consistent with the Z-burst model, application of Occam's razor argues against it. One should regard the work presented here as an interesting possibility at the time, which has not stood the test of time.

Acknowledgments

I wish to thank my collaborators, Dr Matt Garbutt, Dr Terry Goldman, and Professor Jerry Stephenson, for their contribution to the work described, and the Australian Research Council for their support.

References

[1] B. H. J. McKellar, M. Garbutt, G. J. Stephenson, and J. T. Goldman, *Neutrino clustering and the Z-burst model*, arXiv:hep-ph/0106123. Contributed to *20th International Symposium on Lepton and Photon Interactions at High Energies* (LP 01), Rome, Italy, 23-28 July, 2001, and to *International Europhysics Conference on High Energy Physics* (HEP 2001), Budapest, Hungary, 12-18 July, 2001.

[2] M. Garbutt, *From the highest energy neutrinos to the lowest — a study in neutrino phenomenology*. PhD Thesis, University of Melbourne, 2002. This thesis is a source for details of calculations omitted from this report.

[3] K. Greisen, *Progress in Cosmic Ray Physics*, **3**, 1 (1956).

[4] J. Linsley, L. Scarsi, and B. Rossi, *Phys. Rev. Lett.***6**, 485 (1961).

[5] A. A. Penzias and R. W. Wilson, *Astrophys. J.* **142**, 419 (1965).

[6] K. Greisen, *Phys. Rev. Lett.* **16**, 748 (1966).

[7] G. T. Zatsepin and V. A. Kuzmin, *JETP Lett.* **4**, 78 (1966); *Pisma Zh. Eksp. Teor. Fiz.* **4**, 114 (1966).

[8] T. Stanev, R. Engel, A. Mucke, R. J. Protheroe, and J. P. Rachen, *Phys. Rev. D* **62**, 093005 (2000).

[9] R. J. Protheroe and P. A. Johnson, *Astropart. Phys.* **4**, 253 (1996).

[10] J. N. Bahcall and E. Waxman, *Phys. Lett. B* **556**, 1 (2003).

[11] R. J. Protheroe, arXiv:astro-ph/9812055.

[12] A. M. Hillas, *Ann. Rev. Astron. Astrophys.* **22**, 425 (1984).

[13] A. M. Hillas, *Nucl. Phys. Proc. Suppl.* **75A**, 109 (1999).

[14] P. Bhattacharjee and G. Sigl, *Phys. Rept.* **327**, 109 (2000).

[15] J. R. Ellis, N. E. Mavromatos, and D. V. Nanopoulos, *Phys. Rev. D* **63**, 124025 (2001).

[16] S. R. Coleman and S. L. Glashow, *Phys. Rev. D* **59**, 116008 (1999).

[17] T. J. Weiler, *Astropart. Phys.* **11**, 303 (1999).

[18] T. J. Weiler, arXiv:hep-ph/9910316.

[19] K. Mannheim, R. J. Protheroe, and J. P. Rachen, *Phys. Rev. D* **63**, 023003 (2001).

[20] E. Waxman and J. N. Bahcall, *Phys. Rev. D* **59**, 023002 (1999).

[21] G. J. Stephenson, J. T. Goldman, and B. H. J. McKellar, *Int. J. Mod. Phys. A* **13**, 2765 (1998).

[22] X. G. He, B. H. J. McKellar, and G. J. Stephenson, *Phys. Lett.* **B444**, 75 (1998); Erratum- *ibid.* **581**, 270 (2004).

[23] J. Abraham *et al.* [Pierre Auger Collaboration], *Phys. Rev. Lett.* **101**, 061101 (2008).

[24] J. Abraham *et al.* [Pierre Auger Collaboration], *Astropart. Phys.* **29**, 188 (2008); Erratum- *ibid.* **30**, 45 (2008).
[25] T. P. A. Collaboration, arXiv:0903.1127 [astro-ph.HE].

Chapter 11

Neutrons from the Galactic Centre

Raymond R. Volkas

School of Physics, Research Centre for High Energy Physics
The University of Melbourne, Victoria 3010 Australia
E-mail: r.volkas@physics.unimelb.edu.au

The cosmic ray anisotropies observed by AGASA and SUGAR from the direction of the Galactic centre may be explained by extremely high-energy neutron production. Diffusive shock acceleration may accelerate protons to the required high energies, with subsequent charge-exchange $pp \to nX$ collisions then producing the neutrons. The Galactic centre gamma ray detections by several experiments provide strong corroborating evidence for this mechanism, with the photons arising from the decay of neutral pions produced through the collisions of the same shock-accelerated proton population. This scenario may be tested in the future through direct detection of the neutrons by AUGER, and indirectly by Galactic centre neutrino detection by neutrino telescopes such as IceCube. The acceleration site may be associated with the supernova remnant Sagittarius A East, but the purely phenomenological analysis to be presented does not require this identification. Some inconsistencies in the data are discussed.

11.1. Introduction

The origin of cosmic rays (CRs) remains a very important unsolved problem in astrophysics. Cosmic rays are energetic protons and other nuclei that impinge on the Earth, having a largely isotropic flux that obeys a broken power-law over many decades in energy. Above around 10^{15} eV, the spectrum becomes steeper, giving the feature known as the 'knee'. Just below 10^{19} eV, the spectrum subsequently flattens to produce the 'ankle'. At higher energies still, one expects to observe a cut-off in the spectrum because of energy loss due to cosmic ray collisions with the cosmic microwave background photons. The existence of this Greisen-Zatsepin-

Kuzmin (GZK) [1] cut-off is at present controversial due to divergent observations by the AGASA [2] and HiRes [3] experiments. The topic of this paper, however, concerns some interesting physics in the $10^{18} - 10^{18.5}$ eV range, which is below the ankle and well below the putative GZK cut-off.

The most plausible candidate mechanism for accelerating cosmic rays to energies up to, possibly, the ankle is diffusive shock acceleration at strong shocks produced by Galactic supernova remnants (SNRs). The total energetics of the cosmic ray population supports this origin, as does the power-law form of the cosmic ray spectrum. However, specific acceleration sites have yet to be observationally established, and it has been argued that this mechanism struggles to accelerate protons beyond the knee.[a] The identification of acceleration sites is hampered by the fact that the electrically-charged cosmic ray particles are so scattered by the Galactic magnetic field – which is known to have a turbulent as well as a regular component – that they manifest as an isotropic flux at Earth for energies below the ankle. (The ankle itself is plausibly argued to be due to an extragalactic flux emerging from the Galactic-CR 'background' – we shall not be concerned with extragalactic CRs in this work.)

To pinpoint specific CR sources, it would thus clearly be helpful to use electrically-neutral rather than charged particles. The three obvious candidates are photons, neutrinos and neutrons. All three are (actually or potentially) important probes.

The importance of photons is obvious, and indeed 'electromagnetic astronomy' is making great strides in probing ever more energetic astrophysical processes through gamma-ray astronomy. The existing instruments WHIPPLE, CANGAROO and more recently HESS are providing very important data, with the EGRET detector aboard the now-defunct Compton GRO satellite having made seminal contributions in the past. In the relatively near future, we have GLAST to look forward to. There are, however, complications in relying entirely on very high-energy photons to identify CR acceleration sites. First, photons can be attenuated in flight. Second, no unique microscopic process produces such photons. Consider, for example, decay photons from neutral pions produced in proton-proton collisions, and Compton up-scattered photons produced by a population of accelerated electrons. The former are associated with CR acceleration, but the latter are not. The clear identification of 'pion bumps' in existing data has

---

[a]We will discuss a possible way around this impasse later on.

proven to be non-trivial, though by no means out of the question.[b] Finally, even if the photon does have its origin in pion decay, its energy might be down by an order of magnitude or more over the primary proton whose collision gave rise to the pion in the first place.

Neutrinos are also interesting carriers. They arise from the decay of charged pions produced in hadronic collisions, so their discovery would point to the hadronic acceleration required to produce CRs. But from a flux and hence event-rate point of view, the nascent field of neutrino astronomy will be relevant for the CR origin question at lower rather than higher energies, simply because the CR spectrum decreases rapidly with energy.

Finally, there are neutrons. They are, in a way, complementary to neutrinos. Being unstable, lower energy neutrons will mostly decay before reaching the Earth from any given source, with the decay protons subsequently becoming scrambled in direction by the Galactic magnetic field. How scrambled depends, of course, on the source-Earth distance as well as the neutron energy. The main theme of this talk relies on the simple fact that neutrons produced at the Galactic centre (GC) with energies above 10^{18} eV will, on average, traverse the GC-Earth distance before decaying. With the CR-associated neutrino flux at these energies way too small for detection, and in the absence of a gamma-ray observatory sensitive to such hard photons, neutrons emerge as uniquely useful messengers from the sites of CR acceleration to near ankle-energies. Interestingly, some data already suggest neutron emission from the GC region, the topic we review next.

11.2. The AGASA and SUGAR cosmic ray anisotropies

The AGASA cosmic ray detector has observed what appears to be a statistically significant overabundance of CRs with energies in the range $10^{17.9} - 10^{18.4}$ eV from the general direction of the GC [5]. The anisotropy is about 4% in amplitude, and 20° in size. It is accompanied by a similar, though less statistically significant, excess from the direction of Cygnus. Prompted by this result, Bellido *et al.* reanalysed air shower data collected by the SUGAR instrument that operated near Sydney during 1968-1979 [6]. By examining only those events in the energy range quoted above, they were able to also uncover evidence for an anisotropy. Their data are consistent with a point source located at 7.5° from the GC and 6° from the AGASA

---

[b]In fact for the Galactic centre source we will discuss in this paper, evidence for such a bump was found by Blasi and Melia [4].

maximum. These results are consistent with the Galactic plane enhancement of CR events in the range $10^{17.3} - 10^{18.5}$ eV discovered by HiRes [7].

The AGASA and SUGAR anisotropies are difficult to explain using charged particles, because the Galactic magnetic field would need to be fine-tuned to prevent protons having such energies from becoming completely isotropic. The most robust and natural explanation therefore requires neutral particles, most probably neutrons [5, 8, 9]. This hypothesis is consistent with the turn-on of the anisotropy at about 10^{18} eV, because the relativistic gamma factor would allow neutrons created at the GC to reach Earth before decaying from about this energy upward.

As pointed out by Bossa et al. [10], several other features of the data also find natural explanations. One is the morphology of the AGASA signal. The GC is actually out of the field-of-view of AGASA, so they are literally not in a position to directly detect a GC neutron point source. However, some of the neutrons from such a source will decay en route despite the large gamma factor, with the subsequent bending of the decay protons by the Galactic magnetic field transporting them over AGASA's horizon to manifest as a somewhat diffuse 'proton sunset' (using poetic license). The second feature is the cessation of the anisotropy above few$\times 10^{18}$ eV: fewer neutrons decay before reaching the Earth, and for those that do decay no plausible Galactic magnetic field can bend the final state protons over the AGASA horizon at those higher energies. Bossa et al. performed detailed modelling of the trajectories of decay protons in the Galactic magnetic field to justify the above conclusions.[c] The required flux of neutrons in the energy range $10^{17.9} - 10^{18.5}$ eV was found to be about

$$2 \times 10^{-17} \text{ cm}^{-2} \text{ s}^{-1}, \tag{11.1}$$

which is roughly consistent with the SUGAR result of $(9 \pm 3) \times 10^{-18}$ cm^{-2} s^{-1}. The Cygnus anisotropy can be attributed to the magnetically directed motion of protons along a spiral arm. The fact that the SUGAR anisotropy is close to point-like is also readily understood because, unlike AGASA, the GC was within the field of view of this instrument.

To go with these non-trivial self-consistencies there are, however, two problems. The first is that the centroid of the SUGAR excess is not consistent with GC itself, being offset by 7.5°. This would not be a problem if there were a plausible source at the claimed location, a possibility worth serious consideration. A more egregious problem is that while the GC is not

---

[c]Note, however, that the functional form chosen by Bossa et al. for the regular magnetic field component does not obey div$\vec{B} = 0$, so it may be prudent to revisit this issue [11].

in the field-of-view of AGASA, the SUGAR centroid *is*. So, on the face of it, were there to be a new source there, AGASA should have seen it, unless it is variable on the timescale of decades (recall that SUGAR ceased taking data in 1979). It would be premature to adopt a definite resolution for these inconsistencies. For the purposes of this talk, I will simply suppose that a neutron source exists very close to the true GC, leaving the offset re SUGAR as an unsolved problem. The southern hemisphere AUGER detector will hopefully provide new and independent data at the relevant energy scales and perhaps resolve the mystery.

11.3. Sagittarius A east: a possible GC source

A possible source for shock-accelerated protons near the GC is the supernova remnant Sagittarius (Sgr) A East. Evidence supporting this view comes from gamma-ray detections of the GC, including the region occupied by this remnant. The $10^9 - 10^{10}$ eV EGRET source 3EGJ1746-2851 [12] is consistent with this hypothesis, as are TeV-scale air Cerenkov detections by Whipple [13] and, most recently, HESS [14] CANGAROO has also detected the GC [15], but its measured spectrum is much steeper than HESS, so the consistency of those two observations is another problem (which obviously will not be solved in this paper). The simple idea is that shocked protons from Sgr A East collide with ambient protons to produce the neutral pions whose decay photons explain the gamma-ray observations [4, 9, 16, 17]. The fact that the spectral indices for both the EGRET and HESS data are about 2.2 is good evidence in favour of a diffusive shock acceleration origin: shocked protons are distributed according to an $E^{-\gamma}$ power law, with $2 - 2.4$ being plausible values for the spectral index γ, and the decay photons inherit this same value. (Whipple provided a single datum, so it does not contribute to the spectral index determination. Its flux value, however, is consistent with HESS.) The EGRET, Whipple and HESS data are plotted in Fig. 11.1; a figure taken from Ref. [9].

From the plot it is immediately apparent that while the spectral indices from EGRET and HESS agree very nicely, the flux normalisations do not. This means that either one (or both) of these results has a serious normalisation error, or much more plausibly, that the two instruments detected what are effectively *two* sources, even if both are associated with Sgr A East [9]. Being a supernova remnant, Sgr A East is of course not a point-like object, so it is quite possible that different portions of the shock front constitute, phenomenologically speaking, different effective sources. Independent sup-

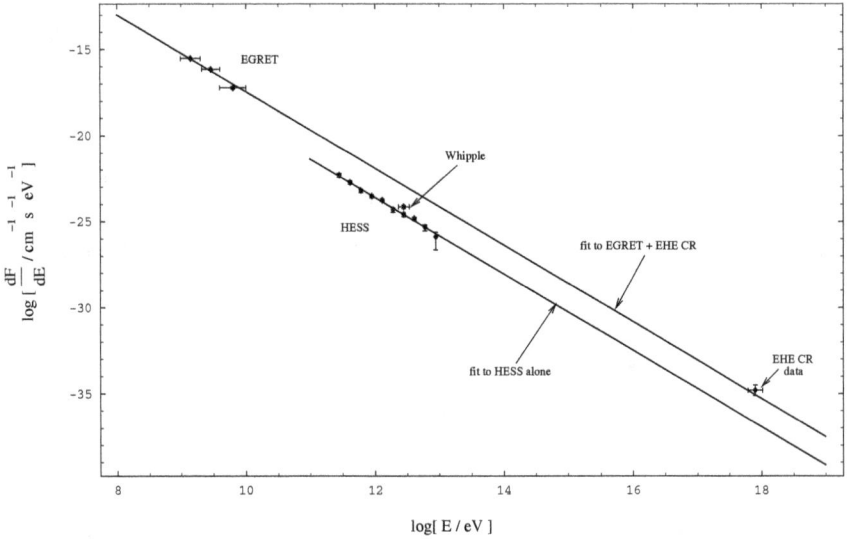

Fig. 11.1.   Figure depicting the EGRET and HESS gamma-ray Galactic centre signals.
The upper solid line arises from a simultaneous fit to EGRET and the neutron flux
deduced from the cosmic ray anisotropies. See text for further explanations.

port for this proposition comes from recent reanalyses of the location of
the EGRET source, which show an offset from the TeV source [18]. The
two-source interpretation requires, of course, that the EGRET signal cuts
off before the first HESS data point at about 3×10^{11} eV.

If the gamma-rays are indeed from neutral pions, then the particle
physics of high-energy proton-proton collisions can be used to predict the
fluxes of other final state particles, especially neutrons from charge exchange
$pp \rightarrow nX$ processes, and neutrinos from the decay of final state charged pions. The normalisations for these other fluxes are set by the gamma-ray
observations. The predictive complementarity between emitted species is a
very important consequence of the hadronic acceleration hypothesis. It is
important to appreciate that the results to be presented below do not rely
on the Sgr A East source identification for their validity [9]. Rather, the
calculation is purely phenomenological: postulate a GC source of accelerated protons with a power law spectrum, then fit to the gamma ray and
anisotropy data. The neutron signal is then a phenomenological prediction.

11.4. Shock acceleration to ankle energies

As mentioned above, it has been argued that while diffusive shock acceleration associated with Galactic SNRs is a most plausible mechanism for accelerating CRs up to the knee, it runs into problems explaining why the spectrum continues beyond this feature (albeit with a slightly steeper spectrum). An important input into this conclusion is the assumption that the shocks are 'parallel', meaning that the regular magnetic field vector is parallel to the shock front normal. In this configuration, one relies on inhomogeneities in the B-field to repeatedly scatter protons and light nuclei across the shock front, thus enabling them to pick up net energy. The highest energy gain will then occur for particles that get scattered back-and-forth across the shock front at the greatest rate, that is, for particles with the smallest mean free path. It is also known that CRs are most strongly scattered by inhomogeneities on the scale of their gyro-radii. Because gyro-radii increase with energy, and there is an upper bound to the scale of the inhomogeneities, an upper limit to the maximum energy to which a particle can be accelerated is obtained. Using B-field and distance scales appropriate to typical SNRs, one can derive that $E_{\max}^{\parallel} \sim \text{few} \times 10^{14} Z$ eV, where Z is the charge of the nucleus [19].

The situation is different for 'perpendicular' shocks, that is, when the B-field is perpendicular to the shock normal (and hence parallel to the shock front) [20]. In this case, the gyro-motion of the charged particles can carry them across the shock front many times between each scattering, thus evading the restriction discussed above. The limiting energy in this case can be computed to be roughly [9],

$$E_{\max}^{\perp} \sim 5 \times 10^{16} Z \left(\frac{B}{20 \ \mu\text{G}} \right) \left(\frac{R_{\text{shock}}}{10 \ \text{pc}} \right) \ \text{eV}, \qquad (11.2)$$

which is clearly an improvement over the parallel case.

However, few$\times 10^{16}$ eV is still a couple of orders of magnitude short of ankle energies. This is where the special conditions of Sgr A East plausibly play a role, because the ambient B-field is expected to be much higher than for typical SNRs. From the equipartition argument, it is known that interstellar magnetic field strengths increase with the square-root of the ambient density. A feature of the GC region is the existence of many molecular clouds, portions of which can have particle concentrations as high as about 10^5 cm^{-3}. Parts of the expanding shell of Sgr A East are indeed colliding with molecular clouds, making it plausible that the magnetic field strengths at those locations are as high as a few milliGauss, far above the typical 20 μG value of other SNRs. When that is fed into Eq. (11.2), a maximum energy of order 10^{19} eV is obtained, so acceleration up to the ankle looks possible [9]. As well as giving rise to a higher B-field, the high target density also explains why Sgr A East is especially luminous in gamma rays, some two orders of magnitude higher than other detected SNR gamma ray emitters [17].

11.5. Neutron flux results

The purely phenomenological calculation [9] begins by postulating a population of accelerated protons obeying a power law,

$$\frac{dn_p}{dE_p}(E_p) \propto E_p^{-\gamma}, \tag{11.3}$$

where n_p is the proton concentration and the spectral index γ should be in the range $2.0 - 2.4$ consistent with diffusive shock acceleration. The constant of proportionality contributes to overall signal strength, so its value is constrained through observational input.

The accelerated protons collide with the ambient protons of a molecular cloud. The pp collisions induce pX and nX final states, where X typically contains many neutral and charged pions. The differential emissivity of a given final state particle f at a given energy E_f is proportional to the product of the target concentration and an integral over proton energy E_p of the proton spectrum multiplied by the differential cross-section with respect to E_p for producing particle f with energy E_f.

Each neutral pion decays to two gamma-ray photons that we will assume provide the higher-energy electromagnetic signals from the GC. The observed gamma-ray flux is then used to normalise the product of the proportionality constant in Eq. (11.3) and the concentration of the target protons. With the normalisation now phenomenologically determined, the neutron flux can be calculated, as a function of the spectral index, based on the known particle physics of the pp reactions. For the neutron calculation, two important parameters are the multiplicity and inelasticity of leading neutrons. The multiplicity is simply the fraction of pp interactions that produce a leading neutron, while the inelasticity is the average fraction of the incident proton energy carried off by the neutron. The adjective 'leading' signifies that the neutron is basically the isospin transform of the incident proton, rather than a soft neutron produced more indirectly (e.g.

through pair production) in the final state. Both of these parameters vary only weakly with energy, and a combination of low-energy experimental data [21, 22] and theoretical extrapolation [23] to the higher energies required for our application suggest that taking them to be approximately constant is sufficient. The values 0.4 for multiplicity [21] and 0.25 for inelasticity [22] have been measured in the lower energy experiments, and we take them to be still valid in the energy regime of concern to us [23]. Our calculations also take into account deviations from scaling for the pp cross-section.

The results are summarised in Figs. 11.1 and 11.2, taken from Ref. [9]. The upper straight line in Fig. 11.1 and the inclined straight line in Fig. 11.2 show fits for two different hypotheses. Figure 11.2 relates to the more likely hypothesis that the TeV-scale HESS and Whipple gamma-ray signals are produced by the same source that produces extremely high energy neutrons. Non-trivially, a good fit to *both* the HESS data and the CR anisotropy datum [the flux quoted in Eq. (11.1)] can be obtained with a spectral index of about 2.0 (the best fit is actually 1.97), consistent with a common origin from shock accelerated protons. The right hand-most datum is the required

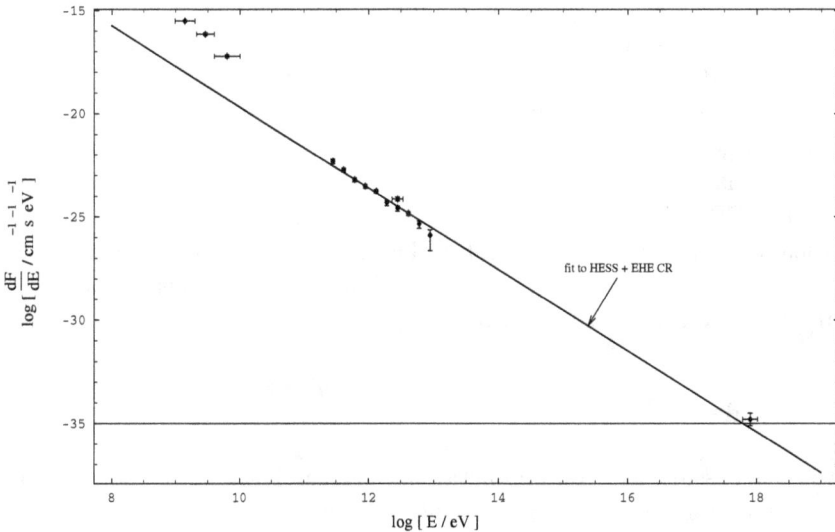

Fig. 11.2. Simultaneous fit to the Galactic centre HESS gamma-ray data and the neutron flux deduced from the cosmic ray anisotropies. See text for further explanations.

neutron flux with errors, and the predicted flux from the fit agrees with it so well that the triangle indicating its location is completed obscured by the experimental data point. (The fact that these *neutron* fluxes are numerically close to the *photon* flux at that energy is an accident.) The EGRET data are above the line, and as already discussed, it is likely they arise from a second effective source that has a cut-off energy below about 10^{11} eV.

For completeness, Fig. 11.1 shows a simultaneous fit to the EGRET and 'neutron' data. An excellent fit is again obtained, this time with $\gamma \simeq 2.23$. This is the less likely hypothesis, however, because it overpredicts the TeV gamma-ray flux by about a factor of 20.

We have also computed the associated neutrino fluxes; see Ref. [24] for the results and discussion.

11.6. Conclusion

The Galactic centre may contribute in an important way to production of extremely high energy cosmic rays. The higher densities and magnetic fields in the region can enable the diffusive shock acceleration mechanism for the perpendicular configuration to produce maximum energies up to 10^{19} eV. The CR anisotropies from the general direction of the GC reported by AGASA and the Bellido *et al.* reanalysis of old SUGAR data are broadly consistent with a neutron point source. These neutrons may be produced from charge-exchange collisions between accelerated and ambient protons. This mechanism can simultaneously explain the GC TeV gamma-ray data and the CR anisotropy. There are also some serious inconsistencies that probably only new data can resolve. These data may come from AUGER, which will see a point source of neutrons from the GC at energies of $10^{18} - 10^{18.5}$ eV if our hypothesis is correct.

Note added:

Another possible neutron production mechanism is the dissociation of nuclei. This was mentioned in Ref. [9] and, after this talk was given, was considered in detail in Ref. [25].

Some time after this talk was given, the Pierre Auger collaboration reported [26] that they did not confirm the anisotropy claimed by AGASA and SUGAR.

Acknowledgments

It was an honour to speak at this celebration of the life and science of Ken Hines, whose passing we mourn. Ken was a teacher, a mentor and a man of great scholarship and culture. He is already sadly missed. I would like to thank my collaborators Roland Crocker, Marco Fatuzzo, Randy Jokipii and Fulvio Melia, and I acknowledge a useful discussion with Matthew Baring. This work was supported by the Australian Research Council.

References

[1] K. Greisen, *Phys. Rev. Lett.* **16**, 748 (1966);
    G. T. Zatsepin and V. A. Kuzmin, *Pis'ma Zh. Eksp.* **4**, 114 (1966) and *JETP Lett.* **4**, 78 (1966).

[2] M. Takeda *et al.*, *Astropart. Phys.* **19**, 447 (2003) and *Phys. Rev. Lett.* **81**, 1163 (1998).

[3] R. U. Abbasi *et al.*, astro-ph/0501317.

[4] P. Blasi and F. Melia, *MNRAS* (submitted, 2003).

[5] N. Hayashida *et al.*, *Astropart. Phys.* **10**, 303 (1999) and astro-ph/9906056.

[6] J. A. Bellido *et al.*, *Astropart. Phys.* **15**, 167 (2001).

[7] D. J. Bird *et al.*, *Ap. J.* **511**, 739 (1999).

[8] L. Jones, *Proc. 21st Int. Cosmic Ray Conf. (Adelaide)*, **2**, 75 (1990).

[9] R. M. Crocker *et al.*, *Ap. J.* **622**, 892 (2005).

[10] M. Bossa, S. Mollerach, and E. Roulet, *J. Phys. G* **29**, 1409 (2003).

[11] J. R. Jokipii, private communication.

[12] H. Mayer-Hasselwander *et al.*, *A.and A.* **335**, 161 (1998).

[13] K. Kosack *et al.*, *Ap. J.* **608**, L97 (2004).

[14] F. A. Aharonian *et al.*, *A. and A.* **425**, L13 (2004).

[15] K. Tsuchiya *et al.*, *Ap. J.* **606**, L115 (2004).

[16] R. M. Crocker, F. Melia, and R. R. Volkas, *Ap. J. S.* **130**, 339 (2000).

[17] F. Melia *et al.*, *Ap. J.* **508**, L65 (1998);
     M. Fatuzzo and F. Melia, *Ap. J.* **596**, 1035 (2003).

[18] D. Hooper and B. Dingus, astro-ph/0212509;
     M. Pohl, astro-ph/0412603.

[19] P. O. Lagage and C. J. Cesarsky, *A. and A.* **125**, 249 (1983).

[20] J. R. Jokipii, *Ap. J.* **255**, 716 (1982) and *Ap. J.* **313**, 842 (1987).

[21] F. T. Dao *et al.*, *Phys. Rev.* **10**, 3588 (1974);
     J. Engler *et al.*, *Nucl. Phys.* **B 84**, 70 (1975);
     W. Flauger and F. Mönning; *Nucl. Phys.* **109**, 347 (1976);
     V. Blobel *et al.*, *Nucl. Phys.* **135**, 379 (1978);
     C. Forti *et al.*, *Phys. Rev. D* **42**, 3668 (1990).

[22] G. M. Frichter, T. K. Gaisser, and T. Stanev, *Phys. Rev. D* **56**, 3135 (1997).

[23] J. N. Capdevielle and R. Attallah, *AIP Conf. Proc.* **276**, *Very High Energy Cosmic Ray Interactions*, ed. L. Jones (Melville: AIP), 448;

J. N. Capdevielle, R. Attallah, and P. Gabinski, *AIP Conf. Proc. 276, Very High Energy Cosmic Ray Interactions*, ed. L. Jones (Melville: AIP), 442.

[24] R. M. Crocker, F. Melia, and R. R. Volkas, *Ap. J.* **622**, L37 (2005).

[25] D. Grasso and L. Maccione, astro-ph/0504323.

[26] Pierre Auger Collaboration (M. Aglietta *et al.*), *Astropart. Phys.* **27**, 244 (2007).

Chapter 12

Quaternions and Octonions in Nature

G. C. Joshi

School of Physics, University of Melbourne
Victoria 3010, Australia
E-mail: girish@unimelb.edu.au

12.1. Introduction

I am honoured to present this in memory of my dear friend Ken.

Quaternions were discovered by Hamilton in 1843 after years of attempts to extend the complex numbers to three dimensions. In those days this subject had a very interesting Australian connection. In 1846 an English lawyer and mathematician Sir James Cockle was appointed chief justice of Queensland. He was a keen follower of Hamilton and his quaternions. During 1889 in a series of fundamental papers the practical applications were realised by Professor Alexander McAulay of the University of Melbourne and later University of Tasmania. He used dual quaternions to describe finite displacement of rigid deformable bodies and also used quaternions as a practical mathematical framework to describe various physical phenomena, including gravitational potentials, Maxwell's equations of electrodynamics and fluid motion.

12.2. Quaternionic quantum mechanics

There are several formulations of quaternionic quantum mechanics (QQM) [1, 2]. An interesting feature of QQM was the formulation of a new kind of gauge symmetry related to the quaternionic phase called "Q covariance". Invariance under this new transformation requires introduction of massive gauge bosons. This work was done by Finkelstein et al. before the Higgs mechanism was introduced. The connection between complex quantum mechanics and QQM continues to be an unresolved problem. However it turns out that there are only two workable ways in which a reliable QQM can be formulated.

- In the first case QQM reduces to complex quantum mechanics at large distances but allows a local structure which is a manifestation of quaternionic structure [3]. Adler used this approach to high energy physics to formulate a model for preon dynamics.
- In the second case the local framework can be complex or quaternionic but the quaternionic make-up is itself dynamic requiring additional gauge fields. This type of approach was used by Brumby and Joshi [4]. Multi-particle correlations in quaternionic quantum systems were studied by Brumby, Joshi and Anderson [5]. A generalised quaternionic quantum mechanics was proposed by Brumby, Hanlon and Joshi to formulate a theory of cosmic strings, and nonbaryonic hot dark matter [6].

Quaternions are isomorphic to Pauli matrices and as noted by Pauli that fundamentality of Pauli matrices lies in the fact that their root is in quaternions. The profound role of SU(2) in physics is discussed by Anderson and Joshi [7] where we conclude that in spite of the remarkable success of SU(2) somehow quaternionic quantum mechanics has not been that successful.

12.3. Octonions supersymmetry and string theory

Octonions were first discovered by Graves in December 1843 and rediscovered by Cayley in 1885. Octonions form a nonassociative algebra. In 1898 Hurwitz showed that there are only four normed division algebras:

- Real numbers
- Complex numbers
- Quaternions
- Octonions

All these numbers satisfy a sum square theorem that is [8],

$$(x_0^2 + x_1^2 + \ldots + x_{n-1}^2)(y_0^2 + y_1^2 + \ldots + y_n^2) = (z_1^2 + \ldots + z_n^2). \qquad (12.1)$$

This equation is satisfied when we define, for $i, j, k \in \{1, n-1\}$ for $n = 1, 2, 4$ and 8

$$x = e_0 x_0 + e_i x_i, \tag{12.2}$$
$$\bar{x} = e_0 x_0 - e_i x_i, \tag{12.3}$$
$$x_0, x_i \in \mathcal{R}, \tag{12.4}$$
$$e_i e_j = -\delta_{ij} + \epsilon_{ijk} e_k, \tag{12.5}$$
$$e_0 e_i = e_i e_0 = e_i, \tag{12.6}$$
$$e_0 e_0 = e_0, \tag{12.7}$$

y and z are defined similarly, and

$$z = xy. \tag{12.8}$$

The case $n = 1$ corresponds to real numbers, when the result is trivial. $n = 2$ gives the complex numbers, in which case the more usual notation is $e_0 = 1, e_1 = i$ and equation (12.1) follows from

$$(x_0^2 + x_1^2)(y_0^2 + y_1^2) =$$
$$(x_0 e_0 + x_1 e_1)(x_0 e_0 + x_1 e_1)^* (y_0 e_0 + y_1 e_1)(y_0 e_0 + y_1 e_1)^*$$
$$= [(x_0 e_0 + x_1 e_1)(y_0 e_0 + y_1 e_1)] [(x_0 + ix_1)(y_0 + iy_1)]^*$$
$$= zz^*$$
$$= (z_0 + iz_1))z_0 + iz_1)^*$$
$$= z_0^2 + z_1^2.$$

Quaternions correspond to $n = 4$, in which case ϵ_{ijk} is the usual Levi-Civita tensor, and octonions correspond to $n = 8$, in which case

$\epsilon_{ijk} = 1$ for $ijk = 123, 145, 176, 246, 257, 347, 365$
and cyclic permutations of these triples,

$\epsilon_{ijk} = -1$ for odd permutations of $ijk = 123, 145, 176, 246, 257, 347, 365$.

The demonstration of the sum square theorem in these cases follows by a similar calculation.

The norm of a division algebra is defined as

$$N(x) = x\bar{x}, \tag{12.9}$$

where $\bar{x}$ is the conjugate, and

$$N(xy) = N(x)N(y), \tag{12.10}$$

and

$$N(x) = x\bar{x} = \bar{x}x = (x_0^2 + \sum_{i=1}^{n-1} x_i^2)e_0. \tag{12.11}$$

Now one can see that the division is defined by

$$x^{-1} = \frac{\bar{x}}{N(x)}. \tag{12.12}$$

The octonionic case is bit more subtle here. Because of nonassociativity we have to define left and right division.

The first application of octonions to physics was pursued by Jordan, Neuman and Wigner. In this formulation observables are represented by octonionic Hermitian matrices and a state is also represented by a Hermitian matrix - a projection operator. The system is quantised through associators rather than commutators. Recently with the advent of supersymmetry and string theory octonions have acquired a very important role. A bimodular representation of ten-dimensional algebra was investigated by Davies and Joshi [9]. It was shown that an octonionic formulation of the fermions leads to an intrinsically ten-dimensional theory. In a series of papers Foot and Joshi [10] investigated the connection between string theory and the division and Jordan algebras. Foot and Joshi studied the Lorentz groups in 3, 4, 6 and 10 dimensions in terms of real, complex, quaternions and octonions. This work required extension of SL(2,C) of the four dimensional Lorentz group. Foot and Joshi also showed that Lorentz groups in 3, 4, 6, and 10 dimensions can be derived from Jordan algebra. Foot and Joshi further showed that super Yang-Mills and superstring theories can be formulated in terms of normed division algebras. Foot and Joshi investigated a deep connection between string theory and the role of division algebras and internal symmetries.

Octonionic gauge theory was investigated by Lassig and Joshi [11]. They studied magnetic charge in nonassociative algebras. They also formulated an octonionic field theory with a magnetic charge, as well as studied a connection between Nambu mechanics and nonassociative algebras. A nonassociative deformation was investigated by Ritz and Joshi [12]. Here we present a precise algebra for the deformed gauge symmetry. This has a novel feature that its associative Lie algebra is preserved but the generators do not close, predicting new physics.

12.4. Octonions and Julia sets

Iterated maps have been extensively studied over real and complex numbers, giving rise to Julia sets. Julia sets for quaternions were investigated by Mandelbrot. Griffin and Joshi [13] extended this work to octonions, the last division algebra.

The simplest map can be written as

$$R \to R^2 + c; R, c \in \mathcal{R}, \tag{12.13}$$

next is the complex map

$$R \to R^2 + c; R, c \in \mathcal{C}, \tag{12.14}$$

then we have quaternionic map

$$R \to R^2 + c + Rc - cR; R, c \in \text{quaternions.} \tag{12.15}$$

We generalised this to octonions by adding a nonassociativity term like:

$$Z \to Z^2 + c + c(Za) - (cZ)a \tag{12.16}$$

where all quantities are octonions.

In this work we have studied the effect of nonassociativity. When nonassociativity is set to zero we get the usual map, when nonassociativity is increased a dramatic phase transition takes place collapsing the Julia set. Figure 12.1[a] shows the (0,1) slice of the map with nonassociativity set to zero. Figure 12.2 shows the (0,1) slice centred at various positions. Figures 12.3 and 12.4 show various slices at different positions with increasing value of nonassociativity. This shows a dramatic effect of nonassociativity in collapsing the map.

In a further study Griffin and Joshi [14] did a detailed study of this phenomena. We found a connection between nonassociativity and the attracting object. Finally Griffin and Joshi [15] also investigated associators in generalised octonionic maps. We showed that structural transitions are a common property of a wide group of octonionic maps. Figure 12.5 is a quadratic map and Fig. 12.6 is an exponential map. A detailed paper by Kricker and Joshi investigated limiting dynamics and bifurcation phenomena in octonionic maps [16]. It exhibits remarkable nonlinear configurations comprising Hopf bifurcations, fixed points, phase locking periodic cycles, tori nontrivial knots, loop doubling and tripling, infinite period doubling cascades and hyperchaos.

[a]Note: all figures are in colour and are found in the Color Figures section, at the end of this chapter.

Acknowledgments

I am grateful to my Ph.D. students Chad Nash, Andrew Davis, Robert
Foot, Brian Hanlon, Chris Griffin, Andrew Kricker, Steve Brumby, and
Chris Lassig with whom this work was done. Part of this work was done in
collaboration with the late Ron Anderson who will always be remembered
by us.

References

[1] D. Finkelstein, J. M. Jauch, and D. Speiser, CERN Report 59-7; reprinted
     in Logico-Algebraic approach to Quantum Mechanics II, (Ed. C. Hooker)
     Reidel, Dordrecht 1979;
     D. Finkelstein, J. M. Jauch, J. M. Schiminovich, and D. Speiser J. Math
     Phys. **3**, 207 (1962);
     D. Speiser J. Math Phys. **4**, 788 (1963).
[2] K. Morita, *Prog. Theor. Phys.* **90**, 219 (1993); *Nuovo Cim. Lett.* **26**, 50
     (1979); *Prog. Theor. Phys.* **65**, 207 (1981); *Prog. Theor. Phys.* **67**, 1860
     (1982).
[3] S. Adler, in *Niels Bohr: Physics and the world: Proceeding of the Niels Bohr
     Centennial Symposium*, Boston, 1985 (Eds. H. Feshbach, T. Matsui, and A.
     Olesan), Harwood Academic, Chur, Switzerland, p. 213, (1988).
[4] S. P. Brumby and Girish C. Joshi, *Chaos Solitons Fractals* **7**, 747 (1996);
     Found. Phys. **26**, 1591 (1996).
[5] S. P. Brumby, Girish C. Joshi , and Ronald Anderson, *Phys. Rev. A* **51**, 976
     (1995).
[6] S. P. Brumby, B. E. Hanlon, and Girish C. Joshi, RCHEP-96-12, *Phys. Lett.*
     B401, 247 (1997).
[7] R. Anderson and G. C. Joshi, *Phys. Essays* **6** 308 (1993); *Chaos Solitons
     and Fractals* **36**, 397 (2008).
[8] L. Sorgsapp and J. Lohmus, *Had J.* **2**, 1388 (1979).
[9] A. J. Davies and Girish C. Joshi, Print-86-1019, University of Melbourne,
     Jul 1986. 13pp. (unpublished); *J. Math. Phys.* **27**, 3036 (1986).
[10] Robert Foot and Girish C. Joshi, *Int. J. Mod. Phys.* **A7**, 3623 (1992); *Int.
     J. Theor. Phys.* **28** 1449 (1989); *Int. J. Theor. Phys.* **28**, 263 (1989); *Lett.
     Math. Phys.* **19**, 65 (1990); *Int. J. Mod. Phys.* **A7** 4395 (1992); *Mod. Phys.
     Lett.* **3A** 999 (1988); *Lett. Math. Phys.* **16**, 77 (1988); *Phys. Rev. D* **37**, 3161
     (1988); *Lett. Math. Phys.* **15**, 237 (1988); *Phys. Lett.* **B199**, 203 (1987); *Mod.
     Phys. Lett.* **A3**, 47 (1988); *Phys. Rev. D* **36**, 1169 (1987).
[11] C. C. Lassig and Girish C. Joshi, *Lett. Math. Phys.* **41** 59 (1997); *Phys. Lett.*
     B400, 295 (1997); *Chaos Solitons and Fractals* **7**, 769 (1996).
[12] A. Ritz and Girish C. Joshi *Chaos Solitons and Fractals* **8**, 835 (1997).
[13] C. J. Griffin and Girish C. Joshi, *Chaos Solitons and Fractals* **2**, 11 (1992).
[14] C. J. Griffin and Girish C. Joshi, *Chaos Solitons and Fractals* **3**, 67 (1993).

[15] C. J. Griffin and Girish C. Joshi, *Chaos Solitons and Fractals* **3**, 307 (1993).

[16] Andrew Kricker and Girish Joshi *Chaos Solitons and Fractals* **5**, 761 (1995).

Colour Figures

Colour figures for Chapter 12 by *G. C. Joshi.*

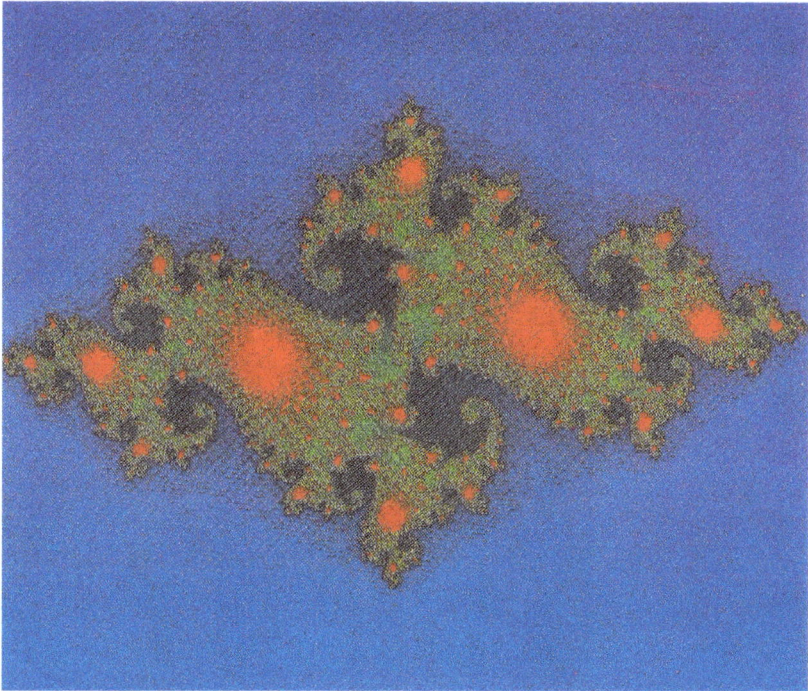

Fig. 12.1.   (0,1) slice of $\mathcal{A}_{\mathcal{R}}(\infty)$, from Ref. [13].

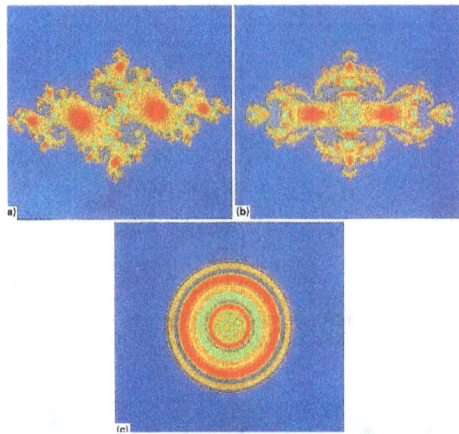

Fig. 12.2. (0,1), (0,2), (1,5) slices of $\mathcal{A}_{\mathcal{R}}(\infty)$, from Ref. [13].

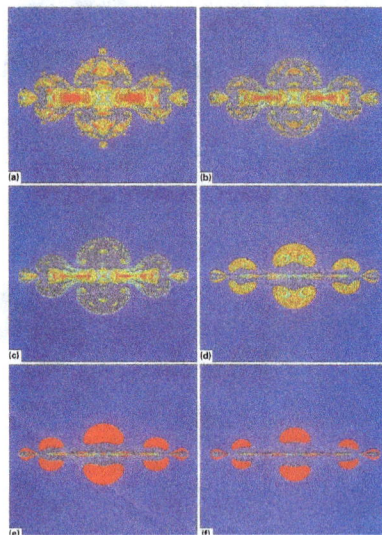

Fig. 12.3. The influence on (0,4) of increasing non-associativity. $\alpha = 0.6$, 1.0, 1.2, 1.95, 2.1, and 2.3, from Ref. [13].

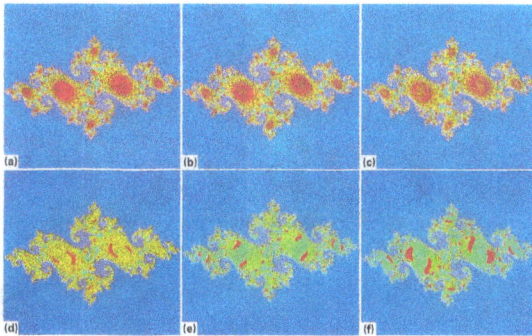

Fig. 12.4. (1,5) (a)-(f) and (5,6) (g)-(l) slices of $\mathcal{A}_{\mathcal{R}}(\infty)$ for $\alpha = 0.6$, 1.0, 1.2, 1.95, 2.1, and 2.3, from Ref. [13].

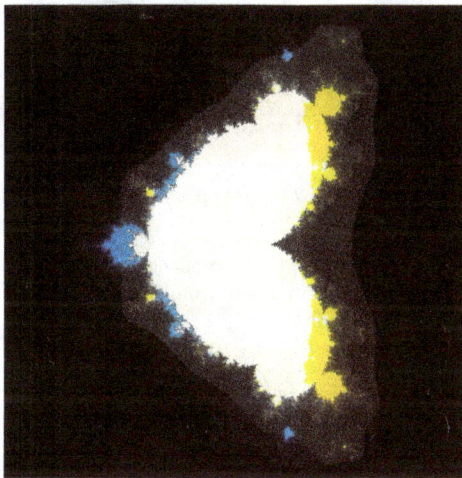

$T(z) = z^4 + z^2$, from Ref. [15].

Fig. 12.5. Large iterate dynamics on the c plane. Colours represent stable regions for different critical points; blue corresponds to the quadratic critical point, and yellow to the quartic critical points, the overlap is rendered white, from Ref. [15].

$$T(z) = \exp(z)$$

Fig. 12.6.   Bifurcation set of the complex exponential map. Each colour represents a unique periodicity of the limit cycle of $z = 0$, from Ref. [15].

Colour figures for Chapter 13 by *F. Melia*.

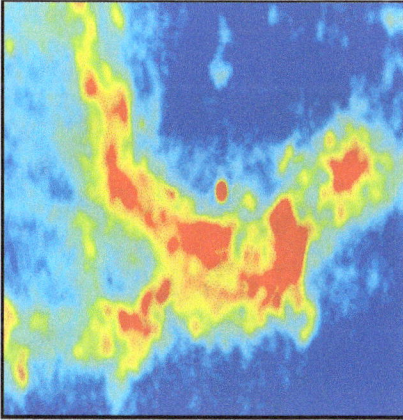

Fig. 13.1.   Sub-arcsec (2 cm) image of Sgr A West and Sagittarius A* (the bright oval object near the middle of the image).

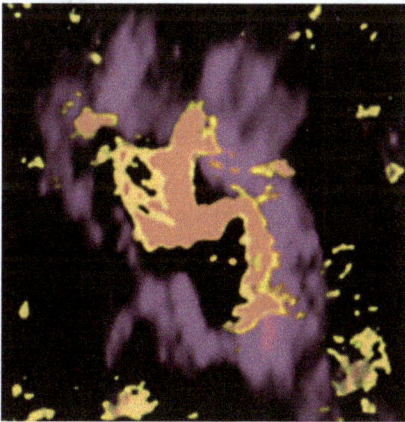

Fig. 13.2.   A radio image of ionised gas (Sgr A West) at $\lambda = 1.2$cm. Most of the ionised gas is distributed in the molecular cavity.

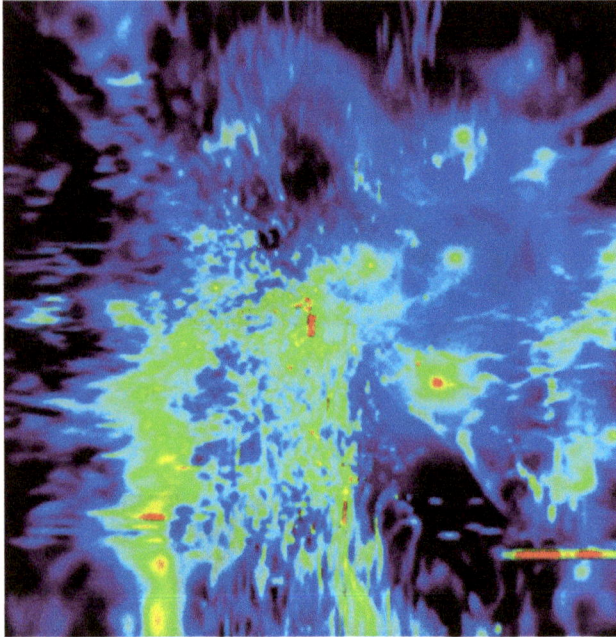

Fig. 13.6. A "snapshot" of the column density (i.e., the gas density integrated along the line of sight) taken at a point in the calculation when the gas distribution had reached stationary equilibrium. Sagittarius A* is in the middle, and the dimensions are approximately 0.5 light years on each side. Some 15 to 20 stars surrounding the black hole each produce an efflux of gas (i.e. "winds"), which collide and form this tessellated pattern of gas condensations, some of which are captured by the black hole and accrete towards it. Several of the wind-producing stars are visible to the right of the image. The colour scale is logarithmic, with red corresponding to a column density of 10^{21} g cm^{-2}, then yellow, blue, and black, which corresponds to 10^{16} g cm^{-2}.

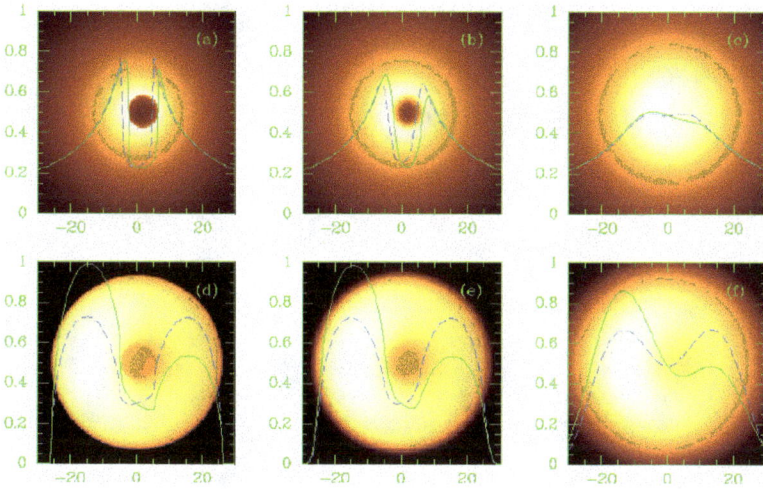

Fig. 13.9. An image of an optically thin emission region surrounding a black hole with the characteristics of Saggitarius A* at the Galactic centre. The black hole is here either maximally rotating ($a_* = 0.998$, panels a-c) or non-rotating ($a_* = 0$, panels d-f). The emitting gas is assumed to be in free fall with an emissivity $\propto r^{-2}$ (panels a-c) or on Keplerian shells (panels d-f) with a uniform emissivity (viewing angle $i = 45°$). Panels a and d show the GR ray-tracing calculations, panels b and e are the images seen by an idealised VLBI array at 0.6 mm wavelength taking interstellar scattering into account, and panels c and f are those for a wavelength of 1.3 mm. The intensity variations along the x-axis (solid green curve) and the y-axis (dashed purple/blue curve) are overlaid. The vertical axes show the intensity of the curves in arbitrary units and the horizontal axes show the distance from the black hole in units of $r_s/2$ which for Saggitarius A* is 3.9×10^{11} cm ~ 3 μas.

Chapter 13

Accretion onto the Supermassive Black Hole at the Centre of Our Galaxy

Fulvio Melia*

Physics Department, the Applied Math Program, and Steward Observatory, The University of Arizona, Tucson, AZ 85721
E-mail: melia@physics.arizona.edu

Towards the end of his life, Ken Hines became fascinated with the physics of accretion onto the black hole at the centre of our Galaxy, a problem well suited to his extensive knowledge of magnetised, high-temperature plasmas. We spent several years formulating a strategy for analyzing the behaviour of ionised gas falling towards Sagittarius A*—as this object is known—but never quite finished the calculations. In this chapter, we will briefly review the circumstances that have made it possible for us to address this problem now, and demonstrate why Ken and I became so excited at the prospect of using the physics of accretion onto Sagittarius A* to probe the spacetime within a mere 20 Schwarzschild radii of its event horizon.

13.1. Introduction

The region bounded by the inner ten lightyears of the Galactic centre contains six principal components that coexist within the central deep gravitational potential well: a supermassive black hole (known as Sagittarius A*), the surrounding cluster of evolved and young stars, a molecular dusty ring, ionised gas streamers, diffuse hot gas, and a powerful supernova-like remnant. Many of the observed phenomena occurring in this complex and unique portion of the Galaxy may be traceable to the interaction between these components.

Though largely shrouded by the intervening gas and dust, the Galactic centre is now actively being explored observationally at radio, sub-millimetre, infrared, X-ray and γ-ray wavelengths with unprecedented

---

*Sir Thomas Lyle Fellow and Miegunyah Fellow, University of Melbourne.

clarity and spectral resolution. And theoretical astrophysicists are attracted to the centre of the Milky Way with the prospect of learning about the physics of black hole accretion and magnetised gas dynamics.

This region is now known to harbour by far the most evident condensation of dark mass, which is apparently coincident with the compact radio source Sagittarius A*. An overwhelming number of observations (proper and radial motion of stars and gas) now strongly supports the view that this object has a mass of at least $2.6 \times 10^6 \ M_\odot$. But the properties of Sagittarius A* are, of course, not independent of its environment. For example, one might naively expect from the observed nearby gas dynamics, that this black hole should be a bright source. Yet it is underluminous at all wavelengths by many orders of magnitude, radiating at only 3×10^{-10} of the theoretical limit for such a mass. Does this imply new accretion physics (as has been proposed) or does it imply something peculiar about Sagittarius A* itself? What now makes asking these questions meaningful is that the extensive sets of data seriously constrain the currently proposed answers.

Over the past decade the number of papers appearing in refereed journals dealing with the theory of phenomena in the Galactic centre, particularly the physics of Sagittarius A*, has doubled roughly every three years. The rate at which papers on the Galactic centre appear is now more than one per week. Our intention here is to briefly summarise the principal observational constraints, and to focus on the key theoretical questions facing the growing number of astrophysicists working in this field.

13.2. The galactic centre environment

The dynamical centre of the Galaxy coincides with Sagittarius A* [1–3], a compact non-thermal radio source no bigger than ~ 1 Astronomical Unit (AU). On a slightly larger scale, the "three-arm" spiral configuration of ionised gas and dust known as Sgr A West [4, 5] engulfs this source in projection as seen in Fig. 13.1, a ~ 6 lightyear $\times$ 6 lightyear image. The cometary-like feature to the north of Sagittarius A* (identified as the bright central spot in this image) is associated with the luminous star IRS 7 [6]. Spectroscopy of the hot gas in the mini-spiral structure seen in Fig. 13.1 [7–9] suggests that it is rotating with a velocity of about 150 km s^{-1} around Sagittarius A* in a counter-clock wise direction. On an even larger scale (~ 9 lightyears), Sgr A West is thought to lie within a large central cavity that is surrounded by a gaseous circumnuclear ring (or circumnuclear disk) [10–13] and is otherwise relatively devoid of neutral gas. A superpo-

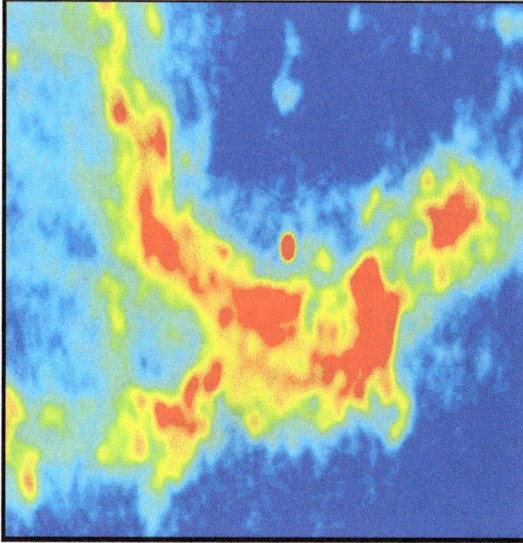

Fig. 13.1.   Sub-arcsec (2 cm) image of Sgr A West and Sagittarius A* (the bright oval object near the middle of the image). See the Colour Figures section for a full colour rendition.

sition of the radio continuum emission from Sgr A West due to free-free radiation with an image showing the distribution of molecular gas is shown in Fig. 13.2. Its three-arm appearance, shown in orange, is superimposed on the distribution of HCN emission which is displayed in red [14]. This suggests that this central cavity is filled with a bath of ultraviolet radiation heating the dust and gas within the inner 24 lightyears of the galaxy [15]. Note that, at the distance to the Galactic centre, the image in Fig. 13.2 corresponds to a size of approximately 12 lightyears on each side [16]. But without doubt, the most imposing structure at the Galactic centre is the central source itself. The prescient application of the then very speculative black hole model for quasars led Lyndon-Bell and Rees [17] to point out that the Galactic centre also should contain a supermassive black hole, perhaps detectable with radio interferometry. Subsequently, Balick and Brown [18] indeed found a compact radio source with the National Radio Astronomy Observatory (NRAO) interferometer at Green Bank and which was later confirmed by Westerbork [19] and by Very Large Baseline Interferometry (VLBI) observations [20]. Eight years after its discovery, the unresolved source was named Sagittarius A* in Ref. [21] to distinguish it from the more extended emission of the Sgr A complex, and to emphasise its uniqueness.

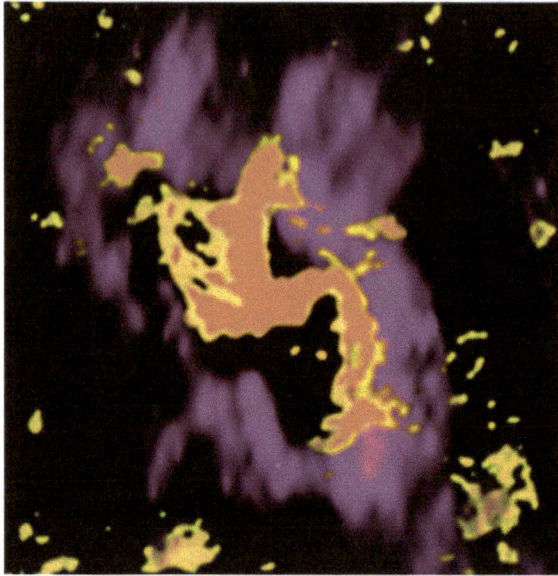

Fig. 13.2.   A radio image of ionised gas (Sgr A West) at $\lambda = 1.2$cm. Most of the ionised gas is distributed in the molecular cavity. See the Colour Figures section for a full colour rendition.

Its radio variability was established also at about this time [22]. The accumulation of these observational signatures make it clear that Sagittarius A* is a very unusual object, rendering it a prime suspect for the location of the putative supermassive black hole. In their review, Genzel and Townes [23] published a diagram (a more recent version of which appears in Fig. 13.3) showing the enclosed mass versus distance from Sagittarius A* with results of later evaluations [24]. These results suggest that there is a concentration of matter with a point-like object (of mass $\sim 3 \times 10^6 M_\odot$) at the Galactic centre. This estimate depended rather sensitively on the mass inferred from the ionised gas motions [7, 25], which some thought could have been influenced by non-gravitational forces (e.g. magnetic fields, stellar winds, etc.). Even so, it was difficult to see how the observed stellar winds and the measured magnetic fields in this general region could be strong enough to produce the observed velocities. In addition, infall from a large distance would have difficulty accounting for the patterns seen [26]. The evidence for the existence of a dark mass concentration has significantly and steadily grown since then, mainly from infra-red observations of stars near Sagittarius A*.

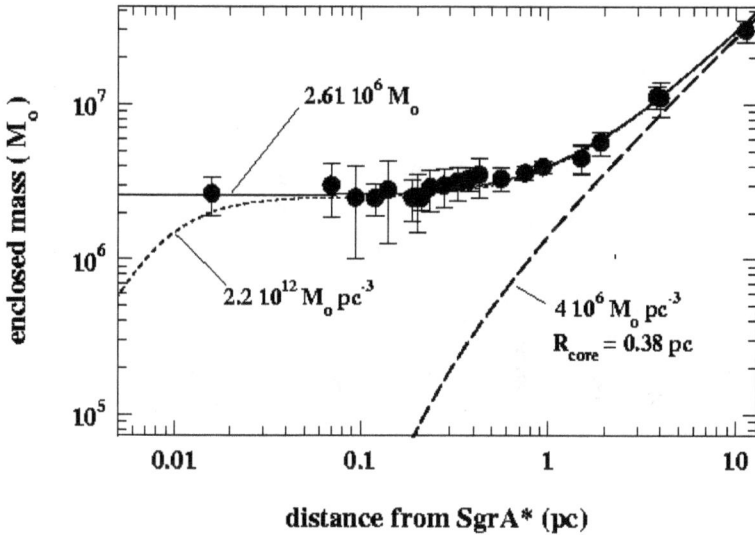

distance from SgrA* (pc)

Fig. 13.3. A plot of the distribution of enclosed mass versus distance from Saggitarius A*. The three curves represent the mass model for a nearly isothermal stellar cluster with a core radius of 1 lightyear (dashed line), the sum of this cluster plus a point mass of $2.61 \pm 0.35 \times 10^6$ $M_\odot$ (solid curve), the same cluster and a dark cluster with a central density of 2.6×10^{11} $M_\odot$ lightyear^{-3} and a core radius of 0.0195 lightyear (dotted curve).

The suggested central dark mass within the inner 0.045 lightyear of the Galactic centre is $2.61 \pm 0.35 \times 10^6$ $M_\odot$.[a] The inferred distribution of matter as a function of distance from Sagittarius A* is shown in Fig. 13.3, and the measured stellar velocity dispersion (shown in the accompanying Fig. 13.5, the data in which were obtained with the Keck telescope [27]), is fully consistent with Keplerian motion about a highly compact central mass concentration. The value of these observations cannot be overstated, since they establish the presence of a dark mass in the Galactic centre beyond a reasonable doubt, even though several systematic uncertainties (on a 10% level) still remain. These uncertainties include that of the exact distance to the Galactic centre (~ 24 thousand lightyears [28]) and the exact mass estimator used to convert velocities to masses. The characteristic size associated with such a mass is the Schwarzschild radius $r_s \equiv 2GM/c^2$, which is here equal to 7.7×10^{11} cm. At a distance of 24 thousand lightyears, this corresponds to 6.4 microarcseconds.

---

[a]0."1 corresponds to 800 Astronomical Units, or roughly 1.2×10^{16} cm at a distance of 24 thousand lightyears.

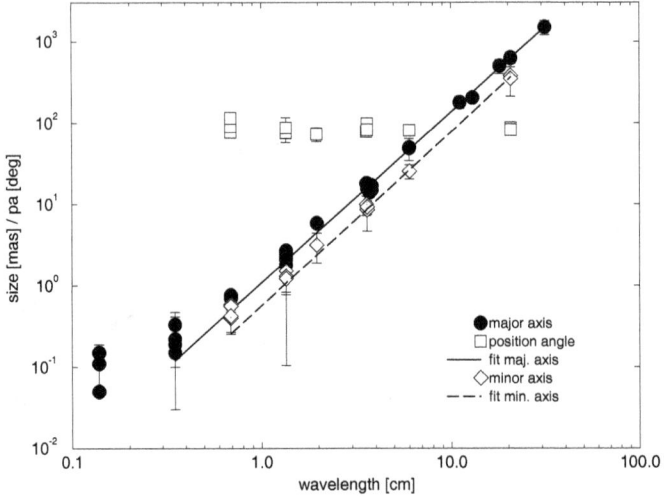

Fig. 13.4.   The major source axis (filled circles) of Saggitarius A*, the minor source axis (open diamonds) and the position angle of the major axis (open squares) as measured by VLBI plotted versus wavelength.

But the main problem in determining the actual size of Sagittarius A* is that its true structure is washed out by scattering in the interstellar medium [29–31], leading to a λ^2 dependence of its diameter as a function of the observed wavelength as shown in Fig. 13.4. Data has been taken from Ref. [32]. The scattering is anisotropic, possibly because of large scale magnetic fields pervading the inner Galaxy [31], with a roughly constant ratio between the major and minor axes of 0.53 at all frequencies below 43 GHz and a constant position angle of 80 ± 3°. The functional form of the scattering size is given as

$$\theta_{minor} = 0.76 \, \text{mas} \, (\lambda/\text{cm})^2 \quad \theta_{major} = 1.42 \, \text{mas} \, (\lambda/\text{cm})^2 \, , \qquad (13.1)$$

and the scattering size apparently has not changed over a decade.

However, it is possible to constrain the mm-to-sub-mm *intrinsic* size of Sagittarius A* to within a factor of 10. This is achieved by observing this object at as high a radio frequency as feasible for which the effects of interstellar scattering are relatively weak. Observations at 86 GHz ($\lambda = 3$mm) and 215 GHz ($\lambda = 1.4$mm) demonstrate that Sagittarius A* is compact on a scale at or below 0.1 mas (1.3×10^{13} cm) for the highest frequencies. This corresponds to ~ 17 Schwarzschild radii for a $2.6 \times 10^6 \, M_\odot$ black hole. While the exact size of Sagittarius A* cannot yet be stated with

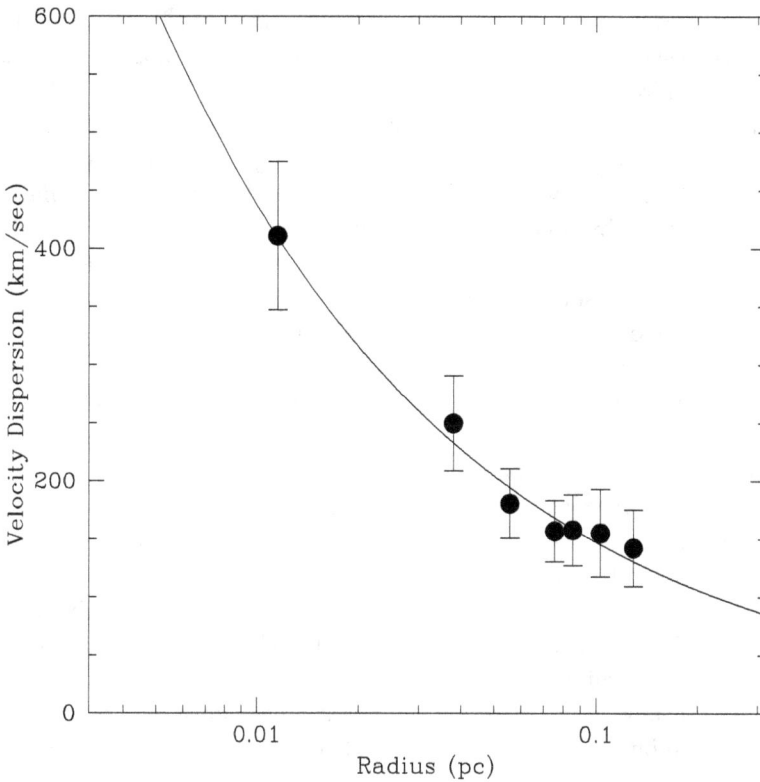

Fig. 13.5. The projected stellar velocity dispersion versus the distance from Sagittarius A*. The solid curve represents Keplerian motion due to a mass concentrated within 0.03 lightyear.

absolute certainty, the latest observations fuel hopes that somewhere in the millimetre wave regime the intrinsic source size will finally dominate over interstellar broadening, allowing a direct comparison with the predictions of various emission theories.

The upper limit to Sagittarius A*'s size ($\sim$ 1 A.U.) requires that its brightness temperature be greater than $\sim 10^{10}$ K for it to produce the power we measure at Earth. This is actually the maximum brightness temperature of incoherent synchrotron emission from an electron plasma. Above this temperature, the radiation is heavily influenced by Compton processes to a frequency well beyond the GHz range. Clearly, though, the emitting particles must reach relativistic energies, i.e. their temperature must exceed $\sim 5 \times 10^9$ K, for them to be efficient synchrotron emitters [33–35], corresponding to electron Lorentz factors in the range of a few up to several hundred.

It is already quite evident, therefore, that in the environment near the black hole we must be dealing with a highly energetic plasma threaded by a magnetic field. But we can go even further in determining the nature of Sagittarius A* by analyzing its variability. In the radio flux density, variations are clearly seen between different epochs. For example, Falcke [36] published the results of 540 daily observations of Saggitarius A* at 2.3 and 8.3 GHz made using the Green Bank Interferometer. The data reveal a high degree of correlation between emission at those frequencies. The lag is apparently less than three days which corresponds to a light travel distance of $\leq 10^{16}$ cm (or roughly 300 A.U.).

More recently, much more rapid variability has been detected in this source at X-ray wavelengths, both with the *Chandra* X-ray satellite and with XMM-*Newton*. X-ray eruptions producing variable emission with a timescale of only 10 minutes suggest that the region within which Sagittarius A* produces its radiative output must be smaller than about one A.U. For a black hole with a mass of 3×10^6 $M_\odot$, this corresponds to a mere 17 Schwarzschild radii from the event horizon. Thus, any attempt to model the physics of accretion onto this object must necessarily contend with the often complex behaviour of relativistic, magnetised plasma orbiting in regions of superstrong gravity.

13.3. Gas dynamics and stellar wind capture

Having described the key observational characteristics of Sagittarius A*, I discuss the physical interpretation of this object. The abundance of gas in the environment surrounding this black hole clearly points to accretion as the incipient cause of its ensuing energetic behaviour. The properties described above are consistent with the idea that Sagittarius A*'s spectrum results from the energy liberated by a compressed hot plasma bound to the central gravitational potential during infall.

Consider the physical state of the gas when the gravitational potential well deepens as the plasma approaches the event horizon. In the classical Bondi-Hoyle (BH) scenario [37], the mass accretion rate for a uniform hypersonic flow is $\dot{M}_{BH} = \pi R_A{}^2 m_H n_w v_w$, in terms of the accretion radius $R_A \equiv 2GM/v_w{}^2$. The conditions at the Galactic centre mean an expected an accretion rate $\dot{M}_{BH} \sim 10^{21}$ g s^{-1} ($\approx 1.6 \times 10^{-5}$ $M_\odot$/yr) onto the black hole, with a capture radius $R_A \sim .06$ lightyear.

In reality the flow past Sagittarius A* is not likely to be uniform, since one might expect many shocks to form as a result of wind-wind collisions within the surrounding medium, even before the plasma reaches R_A. The implications for the spectral characteristics of Saggitarius A*, and thus its nature, are significant.

Simulations of the BH accretion from the spherical winds of a distribution of 10 individual point sources located at an average distance of a few R_A from the central object have been made [38]. The results of these

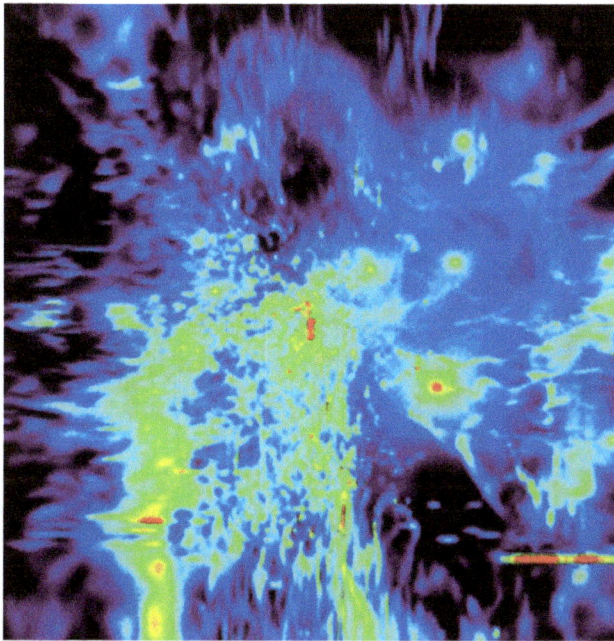

Fig. 13.6. A "snapshot" of the column density (i.e., the gas density integrated along the line of sight) taken at a point in the calculation when the gas distribution had reached stationary equilibrium. Sagittarius A* is in the middle, and the dimensions are approximately 0.5 light years on each side. Some 15 to 20 stars surrounding the black hole each produce an efflux of gas (i.e. "winds"), which collide and form this tessellated pattern of gas condensations, some of which are captured by the black hole and accrete towards it. Several of the wind-producing stars are visible to the right of the image. The colour scale is logarithmic, with red corresponding to a column density of 10^{21} g cm^{-2}, then yellow, blue, and black, which corresponds to 10^{16} g cm^{-2}. See the Colour Figures section for a full colour rendition.

simulations show that the accretion rate depends not only on the distance of the mass-losing star cluster from the accretor but also on the relative spatial distribution of the sources. In addition, the co-existence of hot and warm gas components may itself alter the Bondi-Hoyle capture profile [39], and a capture rate inferred was $\approx 3 \times 10^{-6}$ $M_\odot$/yr. That effect was not included in the simulations [38].

Figure 13.6 is a logarithmic colour scale image of the density profile for a slice running through the centre of the accretor from one of these simulations taken 2,000 years after the winds are "turned on" [38]. Once the stellar winds have cleared the region of the original low density gas, all such simulations point to an overall average density ($\sim 10^3$ cm^{-3}) in agreement with observations.

In Fig. 13.7, the mass accretion rate, $\dot{M}$, and the accreted specific angular momentum, λ are shown versus time, starting 2 crossing times ($\sim$

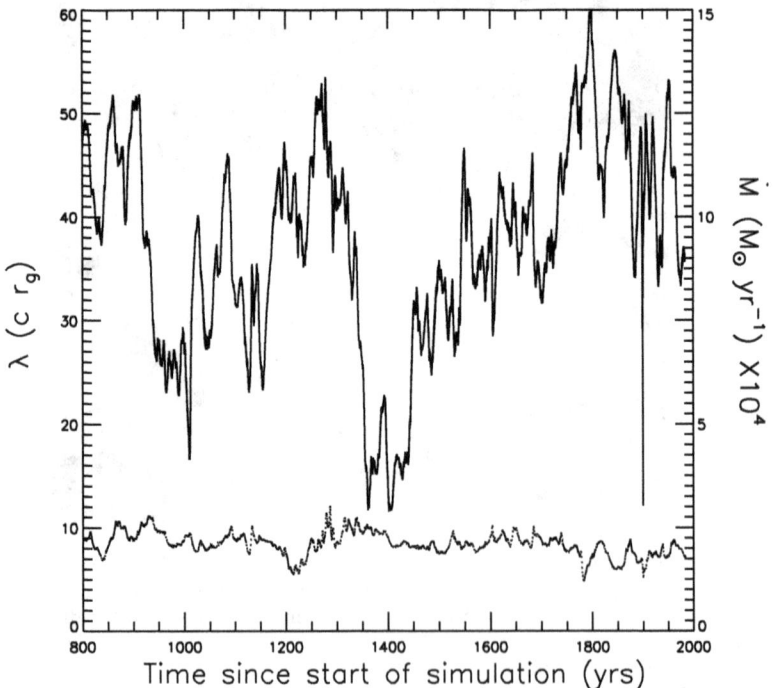

Fig. 13.7.   The upper solid curve is the *magnitude* of the accreted specific angular momentum λ (in units of cr_s). The scale for λ is on the left side while that for $\dot{M}$ is shown on the right. The lower dotted curve is the mass accretion rate $\dot{M}$ (10^{-4} $M_\odot$ yr^{-1}) versus time.

800 years) after the winds are "turned on". λ has units of cr_s, where r_s is the Schwarzschild radius. Results were taken from Ref. [40]. The average value for the mass accretion rate once the system has reached equilibrium is $\dot{M} = 2.1 \pm 0.3 \dot{M}_{BH}$. The mass accretion rate shows high frequency temporal fluctuations (with a period of $\lesssim 0.25$ yr) due to the finite numerical resolution of the simulations. The low frequency aperiodic variations (on the order of 20% in amplitude) reflect the time dependent nature of the flow. Thus, the mass accretion rate onto the central object, and consequently the emission arising from within the accretor boundary, is expected to vary by $\lesssim 20 - 40\%$ (since in some models the luminosity may vary by as much as $\propto \dot{M}^2$) over the corresponding time scale of < 100 years, even though the mass flux from the stellar sources remains constant. The temporal variations in Sagittarius A*'s radio luminosity are probably due, at least in part, to these fluctuations in the accretion rate toward small radii.

Similarly, the accreted λ can vary by 50% over $\lesssim 200$ years with an average equilibrium value of 37 ± 10. It appears that even with a large amount of angular momentum present in the wind, relatively little specific angular momentum is accreted. This is understandable since clumps of gas with a high specific angular momentum do not penetrate to within $1\ R_A$. The variability in the sign of the components of λ suggests that if an accretion disk forms at all, it dissolves, and reforms (perhaps) with a different sense of spin on a time scale of ~ 100 years or less.

The captured gas is highly ionised and magnetised, so it radiates via bremsstrahlung, cyclo-synchrotron and inverse Compton processes. However, the efficiency of converting gravitational energy into radiation is quite small (as little as 10^{-4} in some cases), so most of the dissipated energy is carried inwards [33, 34, 41, 42]. In fact, if the magnetic field is a negligible fraction of its equipartition value (see below), Saggitarius A* would be undetectable at any frequency, except perhaps at soft X-ray energies. But as the plasma continues to compress and fall toward smaller radii, one or more additional things can happen, each of which corresponds to a different theoretical assumption, and therefore a potentially different interpretation.

13.4. Emission by Sagittarius A*

The questions one may ask include the following:

(1) Does the flow carry a large specific angular momentum (in contrast to our expectations from the Bondi-Hoyle simulations) so that it forms a disk with lots of additional dissipation?

(2) Does the flow produce a radiatively dominant non-thermal particle distribution at small radii (e.g. from shock acceleration), or does thermal emission continue to dominate the spectrum?

(3) Does the flow lead to an expulsion of plasma at small radii that forms a non-thermal jet, which itself may then dominate the spectrum?

Attempts to answers these questions, either individually or in combination, have led to a variance of assumptions about the nature of the inflowing gas that then form the basis for the development of different interpretations.

Observationally, one of the key issues is why the infalling gas maintains a low radiative efficiency. In the picture developed by Ref. [43] and updated in Ref. [44] the infalling gas is assumed to carry a very large angular momentum towards the centre, forming a big accretion disk (with an outer edge extending beyond 10^5 Schwarzschild radii or so). The Bondi-Hoyle simulations discussed above suggest that clumps of gas with relatively large specific angular momentum do not penetrate inwards. However, a large disk may form if the viscosity is anomalously high even at large radii. In this case, the overall emission must now include the additional dissipation of the captured angular momentum. To comply with the observed low efficiency of Sagittarius A*, this model therefore also assumes that the electron temperature is much lower than that of the protons ($T_e \ll T_p$). In fact, $T_e < 10^{10}$ K. Since the electrons do the radiating, the efficiency remains small even though the protons are very hot. It is important to point out, in this regard, that the success or otherwise of an advection-dominated model rests on whether or not event horizons really do exist. The low efficiency of such an inflow can be maintained only if the energy transported inward vanishes from view [44].

Large accretion disks such as this are known as ADAFs. Strictly speaking, the acronym ADAF stands for Advection Dominated Accretion Flow, which embraces all forms of accretion (disk or otherwise) in which a large fraction of the dissipated energy is advected inwards by the hot protons, rather than radiated away locally by the electrons. So for example, if the gas flow is quasi-spherical until it gets to within a handful of Schwarzschild radii (as suggested by the Bondi-Hoyle simulations) it may still be advection-dominated if the emissivity of the gas is very low; this may occur when the magnetic field is weak [45, 46]. In practice, however, the term ADAF is conventionally used to denote the category of accretion patterns that involve a large, two-temperature *disk*.

However, such disk models face a significant difficulty, in that there does not appear to be a simple way out of the large dissipation (and consequent radiative efficiency) produced by the wind falling onto the plane [47]. In addition, it is difficult to see how magnetised structures of this size can avoid over-producing synchrotron radiation. A possible resolution to this problem is that the magnetic field within the inflowing gas may be sub-equipartition, which clearly has the effect of lowering the synchrotron emissivity. This effect may be present whether or not the dissipated energy in the flow is advected inwards through the event horizon. The idea that Saggitarius A*'s low emissivity is due to a sub-equipartition magnetic field B deserves close attention, especially in view of the fact that the actual value of B depends strongly on the mechanism of field line annihilation, which is poorly understood. Two processes that have been proposed are (i) the Petschek mechanism [48] in which dissipation of the sheared magnetic field occurs in the form of shock waves surrounding special neutral points in the current sheets and thus, nearly all the dissipated magnetic energy is converted into the magnetic energy carried by the emergent shocks; and (ii) the tearing mode instability [49], which relies on resistive diffusion of the magnetic field and is very sensitive to the physical state of the gas. In either case, the magnetic field dissipation rate is a strong function of the gas temperature and density, so that assuming a fixed ratio of the magnetic field to its equipartition value may not be appropriate.

Kowalenko and Melia [45] used the van Hoven prescription to calculate the magnetic field annihilation rate in a cube of ionised gas being compressed at a rate commensurate with that expected for free-fall velocity onto the nucleus at the Galactic centre. An example of these simulations is shown in Fig. 13.8, for parameter values like those pertaining to the Galactic centre. Whereas the rate of increase $\partial B/\partial t|_f$ in B due to flux conservation depends only on the rate $\dot{r}$ of the gas, the dissipation rate $\partial B/\partial t|_d$ (based on the van Hoven prescription) is a function of the state variables and is therefore not necessarily correlated with $\dot{r}$. Although these attempts at developing a physical model for magnetic field dissipation in converging flows are still rather simplistic, it is apparent from the test simulations that the equipartition assumption is not always a good approximation to the actual state of a magneto-hydrodynamic flow, and very importantly, that the violation of equipartition can vary in degree from large to small radii, in either direction. Coker and Melia [46] have calculated the cm to mm spectrum produced by a quasi-spherical infall in Saggitarius A* using its most recently determined mass, and an empirical fit to the magnetic

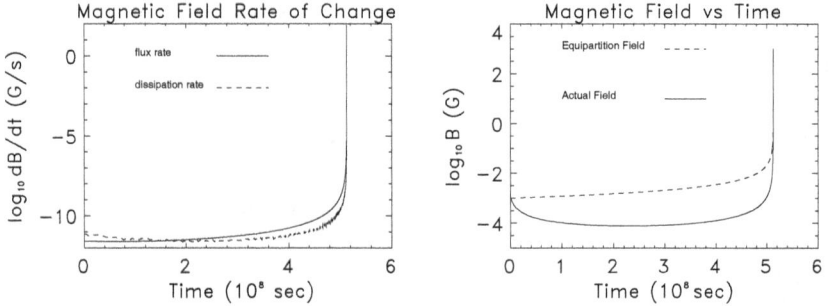

Fig. 13.8.   The *left* panel shows the magnetic field rates of change $\partial B/\partial t|_f$ and $\partial B/\partial t|_d$ due, respectively, to flux conservation and resistive dissipation as functions of time in units of 10^8 seconds. The compression rate is here assumed to be the free-fall velocity at the accretion radius. Solid curve: the rate of increase due to flux conservation; dashed curve: the rate of decrease due to resistive dissipation. The *right* panel shows the magnetic field (solid curve) calculated as a function of time from the rates displayed in left panel. By comparison, the equipartition field B_{eq} is shown here as a dashed curve.

field based on these simulations of magnetic dissipation. Without the additional suppression for the radiative efficiency provided by, for example, a two-temperature flow, the implied magnetic field intensity in Saggitarius A* is limited to values of $5 - 10$ Gauss. Note that the rapid increase in B toward the end of the simulation is associated with the accelerated rate of change in the physical parameters as the gas flows inwards toward a zero radius [45].

It has been suggested [50] that the mm to sub-mm "excess" in the spectrum of Saggitarius A* may be the first indirect evidence for the anticipated circularity of the gas falling into the black hole at $5-25\ r_s$. In their simulation of the Bondi-Hoyle accretion onto Saggitarius A* from the surrounding winds, Coker & Melia (1997) concluded that the accreted specific angular momentum $l \equiv \lambda r_s c$ can vary by 50% over $\lesssim$ 200 years with an average equilibrium value in λ of about 30 or less. The fact that $\lambda \neq 0$ therefore raises the expectation that the plasma must spiral toward smaller radii before flowing through the event horizon. This dichotomy of a quasi-spherical flow at radii beyond $50\ r_s$ and a Keplerian structure toward smaller radii, may be the explanation for Saggitarius A*'s spectrum, including the appearance of the "excess", which is viewed as arising primarily within the circular component [50].

It is essential to self-consistently match the conditions within the Keplerian region of the flow with the quasi-spherical infall further out. Such calculations are necessary and timely. The emission in Saggitarius A* (if produced in the accretion region) requires a very deep potential well, so the case for a massive black hole rather than a distributed dark matter has grown stronger. Whether the radiation mechanism is thermal or non-thermal, the radiative efficiency of the infalling gas must be very low ($< 10^{-5}$). All things considered, this low efficiency is probably due to either a sub-equipartition magnetic field (for either thermal or non-thermal models), or to the separation of the gas into a two-temperature plasma with $T_e \ll T_p$. The current limit on the accreted specific angular momentum appears to be inconsistent with the formation of a large disk, fossil or otherwise [47] favouring instead the circularity of the infalling plasma when it plummets to within $50\,r_s$ of the black hole. The spectral and polarization data pertaining to the sub-mm bump are consistent with this portion of the spectrum arising from the inner Keplerian flow within i $10 - 20\,r_s$ of the accretor.

13.5. Strong gravity effects

The ever growing interest in Saggitarius A* has already yielded a number of tantalising results; the most important being that Saggitarius A* is the best known supermassive black hole candidate. As well as seeking to model the emission from this source, it is worth thinking about using its relative proximity to test the predictions of General Relativity in the strong field limit. For example, the fact that Saggitarius A*'s mass is known so precisely and that the emitting gas is apparently becoming transparent at mm to sub-mm wavelengths near the marginally-stable orbit, means that timing studies of this source with bolometric detectors on single-dish telescopes may reveal the black hole's spin [51].

Surprising as this may seem, we are at the stage where we can begin to ask questions such as "Is there really an event horizon in this source?" The VLBI resolution is rapidly approaching a scale commensurate with the actual size of Saggitarius A*'s event horizon. When we realize that the presence of the sub-mm bump in the spectrum is indicative of a compact emission region a mere couple of Schwarzschild radii in size, it becomes worthwhile exploring the possibility of actually "seeing" the shadow of the black hole using VLBI imaging techniques. This naturally will have to be done at the highest radio frequencies where the resolution is the best, and the scatter-broadening of Saggitarius A* by the intervening medium is the lowest.

At sub-mm wavelengths, the synchrotron emission is not self-absorbed [33, 34]. The medium's transparency at the shortest wavelengths allows a view of the emitting gas all the way down to the event horizon, whose size is $(1 + \sqrt{1 - a_*^2})r_s/2$. Here $r_s \equiv 2GM/c^2$, M is the mass of the black hole, G is Newton's constant, c the speed of light, $a_* \equiv Jc/(GM^2)$ is the dimensionless spin of the black hole in the range 0 to 1, and J is the angular momentum of the black hole. Bardeen [52] described the idealised appearance of a Schwarzschild black hole *in front* of a planar emitting source such as a star, showing that, literally, it would appear as a "black hole" of diameter $\sqrt{27}r_s/2$. At that time, such a calculation was of mere theoretical (rather than practical) interest. To further check whether there is indeed a realistic chance of seeing this "black hole" in the Galactic centre, Falcke, Melia, and Agol [53] simulated the appearance of the emitting gas *surrounding* Saggitarius A* using a general relativistic (GR) ray-tracing code for various combinations of black hole spin, inclination angle, and morphology of the surrounding emission region. The simulations take the scatter broadening and the instrumental resolution of VLBI at sub-mm waves into account.

As revealed by those calculations the presence of an event horizon inside a transparent radiating source will naturally lead to a deficit of photons in the centre, termed [53, 54] a "shadow". The size of the shadow is larger than the event horizon due to the strong bending of light by the black hole and is of order $5r_s$ in diameter. Two disparate cases are reproduced here (see Fig. 13.9), which include a rotating and a non-rotating black hole, a rotating and an inflowing emission region, as well as a centrally peaked and a uniform emissivity.

The shadow can be clearly seen with a diameter of $4.6\,r_s$ (30 μas) in diameter for the rotating black hole and with a radius of $5.2\,r_s$ (33 μas) for the non-rotating case. The emission can be asymmetric due to Doppler shifts associated with rapid rotation (or inflow/outflow) near the black hole. The size of this shadow is within less than a factor of two of the maximum resolution already achieved by sub-mm VLBI ($\sim 50\mu$as [55]). It may also be feasible to do polarimetric imaging at mm and sub-mm wavelengths, which would reveal additional effects of strong gravity distortions [56].

Interestingly, the scattering size of Saggitarius A* and the resolution of global VLBI arrays become comparable to the size of the shadow at a wavelength of about 1.3 mm. As one can see from Fig. 13.9, the shadow is still almost completely washed out for VLBI observations at 1.3 mm, while it is very apparent at a factor two shorter wavelength (Fig. 13.9 panels b

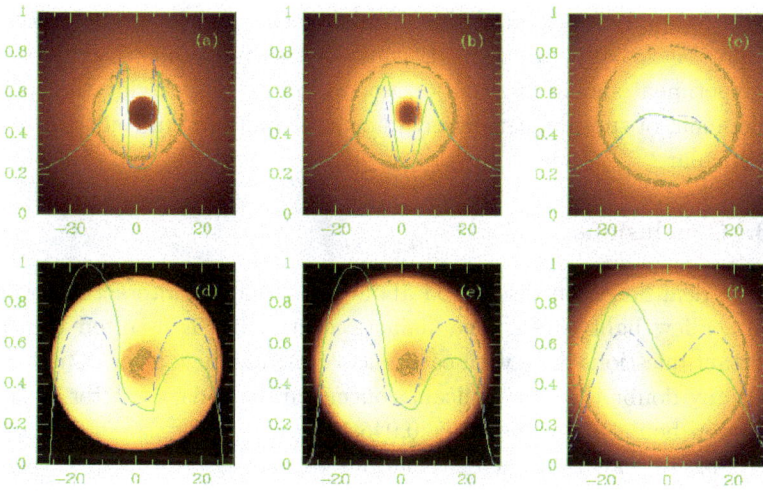

Fig. 13.9. An image of an optically thin emission region surrounding a black hole with the characteristics of Saggitarius A* at the Galactic centre. The black hole is here either maximally rotating ($a_* = 0.998$, panels a-c) or non-rotating ($a_* = 0$, panels d-f). The emitting gas is assumed to be in free fall with an emissivity $\propto r^{-2}$ (panels a-c) or on Keplerian shells (panels d-f) with a uniform emissivity (viewing angle $i = 45°$). Panels a and d show the GR ray-tracing calculations, panels b and e are the images seen by an idealised VLBI array at 0.6 mm wavelength taking interstellar scattering into account, and panels c and f are those for a wavelength of 1.3 mm. The intensity variations along the x-axis (solid green curve) and the y-axis (dashed purple/blue curve) are overlaid. The vertical axes show the intensity of the curves in arbitrary units and the horizontal axes show the distance from the black hole in units of $r_s/2$ which for Saggitarius A* is 3.9×10^{11} cm ~ 3 μas. See the Colour Figures section for a full colour rendition.

and e). In fact, already at 0.8 mm (not shown here) the shadow can be seen easily. Under conditions, such as a very homogeneous emitting region, the shadow would be visible even at 1.3 mm. The technical methods to achieve such a resolution at wavelengths shorter than 1.3 mm are currently being developed and a first detection of Saggitarius A* at 1.4 mm with VLBI has already been reported. Pushing the VLBI technology even further, towards $\lambda = 0.8$ or even $\lambda = 0.6$ mm, should eventually provide the first direct evidence for the existence of an event horizon. Alternatively one could think of space-based X-ray imaging of this shadow as has been proposed recently [57]. However, this technology is still far in the future and Saggitarius A* is a rather weak X-ray source.

The imaging of this shadow would confirm the widely held belief that most of the dark mass concentration in the nuclei of galaxies such as ours is contained within a single object. A non-detection with sufficiently developed techniques, on the other hand, might pose a major problem for the standard black hole paradigm. Because of this fundamental importance, this experiment should be a major motivation for intensifying the current development of sub-mm astronomy in general and mm- and sub-mm VLBI in particular.

13.6. Conclusions

A great deal has been discovered about the principal interactions within the inner few lightyears at the Galactic centre, but as is often the case, important questions arise with each uncovering of a new layer. There is no longer any doubt that a significant concentration of dark matter occupies the region bounded by the inner 0.045 lightyear. This size is sufficiently small that we can rule out distributions of stellar-sized objects, such as neutron stars or brown dwarfs as the constituents. Such a distribution would need to be highly peaked in the centre, and therefore considerably out of equilibrium [58]. Its lifetime would be of order 10^7 years, much smaller than the age of the galaxy [59], leaving us to ponder why we are viewing this region at such a special time. That there is a massive point-like object in the middle is now hard to dispute. It doesn't move relative to objects around it, and it has a spectrum like no other in the Milky Way, though it shares many characteristics in common with the cores of other nearby galaxies.

One of the principal problems now facing us is to understand how in fact Saggitarius A* produces its spectrum. The Galactic centre is rich in gas, and some of it must be funnelling into the black hole. Yet this process does not appear to be converting very much kinetic and gravitational energy into radiation, making Saggitarius A* extremely sub-Eddington. This departure from our naive expectations is forcing us to rethink the basic elements of accretion physics. So theorists are now grappling with questions such as:

(1) is the inflow advection dominated, carrying most of its energy through the event horizon?
(2) is the assumption of equipartition between the magnetic field and the gas an over-simplification that leads to a great overestimation of the magnetic field intensity, and hence of the synchrotron emissivity?
(3) does the plasma separate into two temperatures as it gets compressed and heated?
(4) does the black hole and/or the infalling plasma produce a jet at small radii that then dominates the emissivity from this source?

Perhaps one of the most exciting developments in this program will be the imaging of Saggitarius A*'s shadow against the backdrop of optically thin emitting plasma at sub-mm wavelengths within the next 5 to 10 years. The appearance of this shadow is a firm prediction of General Relativity, which mandates a unique shape and size for the region where light bending and capture are important. There has never been such an opportunity to place the existence of black holes on such a firm footing. Galactic black hole binaries contain compact objects that are too small, and the cores of other galaxies are simply too far away. Saggitarius A* at the Galactic centre has a size that is now on the verge of detection with sub-mm VLBI. This coming decade may finally give us a view into one of the most important and intriguing predictions of General Relativity.

References

[1] A. Eckart, R. Genzel, R. Hofmann, B. J. Sams, and L. E. Tacconi-Garman, *Astrophys. J.* **445**, L23 (1995).

[2] K. M. Menten, M. J. Reid, A. Eckart, and R. Genzel, *Astrophys. J.* **475**, L111 (1997).

[3] A. M. Ghez, B. L. Klein, M. Morris, and E. E. Becklin, *Astrophys. J.* **509**, 678 (1998).

[4] R. D. Ekers, J. H. van Gorkom, U. J. Schwarz, and W. M. Goss, *Astro. Ap.* **122**, 143 (1983).

[5] K. Y. Lo and M. J. Claussen, *Nature* **306**, 647 (1983).

[6] F. Yusef-Zadeh and M. Wardle, *Astrophys. J.* **405**, 584 (1993).

[7] E. Serabyn, J. H. Lacy. C. H. Townes, and R. Bharat, *Astrophys. J.* **326**, 171 (1988).

[8] T. M. Herbst, S. V. W. Beckwith, W. J. Forrest, and J. L. Pipher, *Astrophys. L.* **105**, 956 (1993).

[9] D. A. Roberts, F. Yusef-Zadeh, and W. M. Goss, *Astrophys. J.* **459**, 627 (1996).

[10] E. E. Becklin, I. Gatley, and M. W. Werner, *Astrophys. J.* **258**, 135 (1982).

[11] J. A. Davidson, M. W. Werner, X. Wu, F. F. Lester, P. M. Harvey, M. Joy, and M. Morris, *Astrophys. J.* **387**, 189 (1982).

[12] H. M. Latvakoski, G. J. Stacey, G. E. Gull, and T. L. Hayward, *Astrophys. J.* **511**, 761 (1999).

[13] R. Zylka, R, Güsten, S. Philipp, H. Ungerechts, P. G. Mezger, and W. J. Duschl, in ASP Conf. Ser. 186: *The Central Parsecs of the Galaxy*, eds. H. Falcke, A. Cotera, W. Duschl, F. Melia, and M. J. Rieke (San Francisco: Astronomical Society of the Pacific), 415 (1999).

[14] M. C. H. Wright and D. C. Backer, *Astrophys. J.*, **417**, 560 (1993).

[15] F. Yusef-Zadeh, S. R. Stolovy, M. Burton, M. Wardle, F. Melia, T. J. W. Lazio, N. E. Kassim, and D. A. Roberts, in ASP Conf. Ser. 186: *The Central*

    Parsecs of the Galaxy, eds. H. Falcke, A. Cotera, W. Duschl, F. Melia,
    and M. J. Rieke (San Francisco: Astronomical Society of the Pacific), 197
    (1999).

[16] F. Yusef-Zadeh, F. Melia, and M. Wardle, *Science*, **287**, 85 (2000).

[17] D. Lynden-Bell and M. J. Rees, *MNRAS* **152**, 461 (1971).

[18] B. Balick and R. L. Brown, *Astrophys. J.* **194**, 265 (1974).

[19] R. D. Ekers, W. M. Goss, U. J. Schwarz, D. Downes, and D. H. Rogstad,
    Astro. Ap. **43**, 159 (1975).

[20] K. Y. Lo, R. T. Schilizzi, M. H. Cohen, and H. N. Ross, *Astrophys. J.* **202**,
    L63 (1975).

[21] R. L. Brown, *Astrophys. J.* **262**, 110 (1982).

[22] R. L. Brown and K. Y. Lo, *Astrophys. J.* **253**, 108 (1982).

[23] R. Genzel and C. H. Townes, *Ann. Rev. Ast. and Astrophys.* **25**, 377 (1987).

[24] R. Genzel and A. Eckart, in ASP Conf. Ser. 186: *The Central Parsecs of
    the Galaxy*, eds. H. Falcke, A. Cotera, W. Duschl, F. Melia, and M. J. Rieke
    (San Francisco: Astronomical Society of the Pacific), 3 (1999).

[25] E. Serabyn and J. H. Lacy, *Astrophys. J.* **293**, 445 (1985).

[26] C. H. Townes, in IAU Symp. 169: *Unsolved Problems of the Milky Way*,
    169, 149 (1996).

[27] A. M. Ghez, M. Morris, and E. E. Becklin, in ASP Conf. Ser. 186: *The Cen-
    tral Parsecs of the Galaxy*, eds. H. Falcke, A. Cotera, W. Duschl, F. Melia,
    and M. J. Rieke (San Francisco: Astronomical Society of the Pacific), 18
    (1999).

[28] M. J. Reid, *Ann. Rev. Ast. and Astrophys.* **31**, 345 (1993).

[29] R. D. Davies, D. Walsh, and R. S. Booth, *MNRAS* **177**, 319 (1976).

[30] H. J. van Langevelde, D. A. Frail, J. M. Cordes, and P. J. Diamond, *As-
    trophys. J.* **396**, 686 (1992).

[31] F. Yusef-Zadeh, W. Cotton, M. Wardle, F. Melia, and D. A. Roberts,
    Astrophys. J. **434**, L63 (1994).

[32] T. P. Krichbaum, A. Witzel, and J. A. Zensus, in ASP Conf. Ser. 186:
    The Central Parsecs of the Galaxy, eds. H. Falcke, A. Cotera, W. Duschl,
    F. Melia, and M. J. Rieke (San Francisco: Astronomical Society of the
    Pacific), 89 (1999),

[33] F. Melia, *Astrophys. J.* **387**, L25 (1982).

[34] F. Melia, *Astrophys. J.* **426**, 577 (1994).

[35] R. Mahadevan, R. Narayan, and I. Yi, *Astrophys. J.* **465**, 327 (1996).

[36] H. Falcke, in ASP Conf. Ser. 186: *The Central Parsecs of the Galaxy*, eds.
    H. Falcke, A. Cotera, W. Duschl, F. Melia, and M. J. Rieke (San Francisco:
    Astronomical Society of the Pacific), 113 (1999).

[37] H. Bondi and F. Hoyle, *MNRAS* **104**, 273 (1944).

[38] R. F. Coker and F. Melia, *Astrophys. J.* **488**, L149 (1997).

[39] F. Baganoff, *et al.*, *Nature* **413**, 45 (2001)

[40] F. Melia and R. F. Coker, *Astrophys. J.* **511**, 750 (1999).

[41] S. L. Shapiro, *Astrophys. J.* **180**, 531 (1973).

[42] J. R. Ipser and R. H. Price, *Astrophys. J.* **255**, 654 (1982).

[43] R. Narayan, I. Yi, and R. Mahadevan, *Nature* **374**, 623 (1995).

[44] R. Narayan, R. Mahadevan, J. E. Grindlay, R. G. Popham, and C. Gammie, *Astrophys. J.* **492**, 554 (1998).

[45] V. Kowalenko and F. Melia, *MNRAS* **310**, 1053 (1999).

[46] R. F. Coker and F. Melia, *Astrophys. J.* **534**, 723 (2000).

[47] H. Falcke and F. Melia, *Astrophys. J.* **479**, 740 (1997).

[48] H. E. Petschek, *Astrophys. Sp. Sc.* **264**, 9 (1998).

[49] G. van Hoven, D. L. Hendrix, and D. D. Schnack, *J. Geophys. Res.* **100**, 19819 (1995).

[50] F. Melia, S. Liu, and R. F. Coker, *Astrophys. J.* **553**, 146 (2001).

[51] F. Melia, B. Bromley, S. Liu, and C. Walker, *Astrophys. J. Lett.*, **554**, L37 (2001).

[52] J. M. Bardeen, in *Black Holes*, eds. C. DeWitt and B. S. DeWitt (New York: Gordon and Breach), 215 (1973).

[53] H. Falcke, F. Melia, and E. Agol, *Astrophys. J.* **528**, L13 (2000).

[54] A. de Vries, *Class. Quantum Grav.* **17**, 123 (2000).

[55] F. T. Rantakyro, *et al. Astron. Astrophys. Supp.* **131**, 451 (1998).

[56] B. Bromley, F. Melia, and S. Liu, *Astrophys. J. Lett.* **555**, L83 (2001).

[57] W. Cash, A. Shipley, S. Osterman, and M. Jay, *Nature* **407**, 160 (2000).

[58] R. Genzel, A. Eckart, T. Ott, and F. Eisenhauer, *MNRAS* **291**, 219 (1997).

[59] E. Maoz, *Astrophys. J.* **494**, L181 (1998).

www.ingramcontent.com/pod-product-compliance
Lightning Source LLC
Chambersburg PA
CBHW050556190326
41458CB00007B/2072